"Shafie's *The Administrative Presiden* ..... *Environment* is a welcome addition to the literature. It offers keen insight into the exceptional role of the presidency in shaping U.S. environmental policy, and with its focus on the Trump administration, it brings much-needed critical evaluation of how the administrative presidency can reshape core environmental policies even without passage of new legislation."

—*Michael Kraft, Professor Emeritus, Political Science and Public and Environmental Affairs, and Herbert Fisk Johnson Professor of Environmental Studies Emeritus, University of Wisconsin-Green Bay.*

"Never before has the discussion on environmental policy been more important, yet our ability to resolve environmental problems remains seriously limited. *The Administrative Presidency and the Environment* provides a roadmap of how we have reached this point and, by extension, a map forward. Shafie's analysis of the growth of the administrative presidency replacing statutory progress with regulatory efforts, is an important foundation in understanding environmental policy making in the United States. Shafie's detailed assessment of these dynamics is a valuable resource for scholars of environmental policy as well as those for whom this book is a first read on environmental issues."

—*Matthew A. Cahn, Vice Provost, California State University Northridge.*

"David Shafie offers an especially useful analysis of how recent American Presidents from Clinton through Trump have used executive authority to expand or reduce the scope of environmental policymaking. His discussion of substantive policy cases dealing with natural resources, pollution control and climate change reveals that key decisions are increasingly made by chief executives rather than legislators thanks to their ability to apply various approaches such as rulemaking, appointments, discretionary judgments, and budgetary priorities."

—*Charles Davis, Colorado State University.*

THE ADMINISTRATIVE PRESIDENCY AND THE ENVIRONMENT

The growth of the administrative state and legislative gridlock has placed the White House at the center of environmental policymaking. Every recent president has continued the trend of relying upon administrative tools and unilateral actions to either advance or roll back environmental protection policies. From natural resources to climate change and pollution control, presidents have more been willing to test the limits of their authority, and the role of Congress has been one of reacting to presidential initiatives.

In *The Administrative Presidency and the Environment: Policy Leadership and Retrenchment from Clinton to Trump*, David M. Shafie draws upon staff communications, speeches and other primary sources. Key features include detailed case studies in public land management, water quality, toxics, and climate policy, with particular attention to the role of science in decision-making. Finally, he identifies the techniques from previous administrations that made Trump's administrative presidency possible.

Shafie's combination of qualitative analysis and topical case studies offers advanced undergraduate students and researchers alike important insights for understanding the interactions between environmental groups and the executive branch as well as implications for future policymaking.

David M. Shafie is an associate professor in the Department of Political Science at Chapman University.

THE ADMINISTRATIVE PRESIDENCY AND THE ENVIRONMENT

Policy Leadership and Retrenchment from Clinton to Trump

David M. Shafie

Routledge
Taylor & Francis Group

NEW YORK AND LONDON

First published 2020
by Routledge
52 Vanderbilt Avenue, New York, NY 10017

and by Routledge
2 Park Square, Milton Park, Abingdon, Oxon OX14 4RN

Routledge is an imprint of the Taylor & Francis Group, an informa business

© 2020 Taylor & Francis

Library of Congress Cataloging-in-Publication Data
Names: Shafie, David M., author. | Routledge (Firm)
Title: The administrative presidency and the environment : policy leadership and retrenchment from Clinton to Trump / David M. Shafie.
Description: First Edition. | New York : Routledge, 2020. | Includes bibliographical references and index. |
Identifiers: LCCN 2020006240 (print) | LCCN 2020006241 (ebook) |
ISBN 9781138596139 (Hardback) | ISBN 9781138596146 (Paperback) |
ISBN 9780429487927 (eBook) | ISBN 9780429947391 (Adobe PDF) |
ISBN 9780429947377 (mobi) | ISBN 9780429947384 (ePub)
Subjects: LCSH: Presidents--United States--Case studies. | Environmental management--Political aspects--United States--Case studies. |
Environmental policy--United States--Decision making--Case studies. |
Political leadership--United States--Case studies. | Political culture--United States--Case studies. | United States--Politics and government--Case studies. | Executive power--United States. |
Administrative agencies--United States--Management.
Classification: LCC JK511 .S537 2020 (print) | LCC JK511 (ebook) |
DDC 363.7/05610973 223--dcundefined
LC record available at https://lccn.loc.gov/2020006240
LC ebook record available at https://lccn.loc.gov/2020006241

ISBN: 978-1-138-59613-9 (hbk)
ISBN: 978-1-138-59614-6 (pbk)
ISBN: 978-0-429-48792-7 (ebk)

Typeset in Bembo
by Taylor & Francis Books

This book is dedicated to Tara, who will be the beneficiary of the choices we make today.

CONTENTS

ILLUSTRATIONS

Figures

Tables

ACKNOWLEDGEMENTS

A number of people contributed to this book, and it certainly would not have been possible without them. Matthew Cahn was a source of advice and encouragement and helped persuade me that it was an idea worth pursuing. I am especially grateful to Robert Durant and Daniel Fiorino for providing constructive feedback on early versions of the manuscript. I also wish to thank Natalja Mortensen, Charlie Baker, and the staff at Routledge for guiding me through the process.

The contributors include my research assistants in the Henley Social Science Research Lab at Chapman University: Muhammad Karkoutli, Phoebe DeVos-Cole, Tiana-Renee Silva, Brett Robertson and Sara Juarez. Kaitlin Bishop and Christopher Romanowski of the Environmental Science and Public Policy program at Chapman also provided valuable research assistance.

The librarians and staff of the William Clinton Presidential Library helped me locate documentary materials. I received an institutional grant from Chapman University for this archival research, and I am grateful to Deans Patrick Quinn and Menas Kafatos for supporting my travel for research and for sending me to Copenhagen.

Lastly, I want to thank those who made this book possible by reading versions of the manuscript, providing thoughtful criticism and moral support throughout the writing process. In that regard, Ann Gordon, Sarah Gordon and Audrey Gordon each contributed to this project in their own special ways.

1

ENVIRONMENTAL POLICY GRIDLOCK

In his 2018 State of the Union Address, Donald Trump kicked off the second year of his presidency by boasting, "We have ended the war on American energy, and we have ended the war on beautiful, clean coal."[1] His first year was a lucrative one for the coal industry, thanks to a number of regulatory rollbacks requested by a coal baron and political supporter. After donating $300,000 to Trump's inaugural fund, Robert Murray followed up with a memo entitled "Action Plan for the Administration of Donald J. Trump," which detailed 16 specific actions to benefit the industry. These concessions ranged from rescinding regulations on strip mining to withdrawing from the 2015 Paris Climate Accord.

None of Murray's directives included requests for new legislation, and this was by design. Many of the rules he objected to were mandated by federal laws, such as the Clean Air Act and Surface Mining Conservation and Reclamation Act, which remained popular. Congress was reluctant to update and amend these venerable statutes, despite flaws that occasionally undermined their implementation. In this climate of political conflict created by unclear and outdated federal statutes, the executive branch is in a stronger position to lead on environmental issues by influencing policy outcomes at the implementation stage. Thus, the coal baron's wish list consisted of policy changes that were possible to achieve without going to Congress: administrative rule changes, unilateral actions, budget cuts, specific political appointments, and the nomination of conservative judges to the federal bench. Even though Trump, a Republican, was elected with Republican majorities in both houses of Congress, the administrative strategy offered certain advantages. Bypassing Congress, he could pursue his policy goals without the scrutiny that comes with the legislative process. Without the deliberation that entails, he could also expect faster results; by 2018, the administration had acted on all 14 of his agenda items.

The Republican 115th Congress played a supporting role in these regulatory rollbacks, approving one resolution after another under the Congressional Review Act (CRA). However, this was only a minor role in advancing Trump's agenda, which was concerned with reversing his predecessor's initiatives on pollution, energy and climate, in ways that would benefit the fossil fuel industries. Trump's go-it-alone approach was also a departure from the presidents who preceded him. For example, Barack Obama tried working with Congress to enact ambitious climate legislation early in his presidency, but that effort was soon abandoned in favor of an executive-centered policy agenda.

The turning point for Obama came after the Democrats' loss of their majority in the House of Representatives in 2010. The Center for American Progress, a think tank led by former Clinton Chief of Staff John Podesta, released a report and a series of recommendations that the president use administrative tools to enact a progressive agenda before the next election, rather than waste his energy pursuing legislation in a gridlocked Congress.[2] Rulemaking, executive orders, and agency management led the list of policy instruments that Podesta singled out: "The U.S. Constitution and the laws of our nation grant the president significant authority to make and implement policy." The CAP report identified several urgent issue areas, as well as administrative actions that could be implemented by the end of 2012. Some of the 34 recommendations involved using executive authority under recent statutes to overcome compromises made in Congress. For example, Obama could ensure the new consumer financial protection agency could wield strong and independent regulatory authority. He was urged to administratively withdraw federal lands for preservation, and promote preventative health care by actively guiding rules to implement the Affordable Care Act. Other more modest goals might be accomplished in additional varied policy areas, particularly in education, simply by advocating rules that would make college financial aid more transparent. In addition, the president was urged to stop jailing noncriminal immigrants, to create a virtual U.S. statistical agency, and to order that solar panels be placed on the roofs of Air Force hangars.[3]

Even in Obama's first few months in office, when many environmentalists were elated to see a president hold the first-ever "green" inauguration and publicly support their causes, the political landscape barely resembled the more verdant environmental era of the 1970s. During that fruitful decade, activists pressured Congress to pass a series of sweeping environmental laws, and the environmental movement enjoyed its greatest policy successes and more favorable public opinion. The new landscape offered bleak prospects for Obama to persuade Congress to enact such significant environmental legislation. His administration pursued ENR priorities without Congress as much as possible, relying upon directives and administrative rulemaking to reverse Bush-era policies on pollution, endangered species, and strip mining. This allowed Obama to save political capital to bargain with lawmakers for higher priorities, mainly the economy and health care reform. With the lone exception of a toxic chemical reform bill delivered late in his second term, Obama's environmental policy accomplishments were the result of administrative actions.

There is also a significant downside to an administrative strategy. Policy change by administrative action, compared to an act of Congress, can be fleeting. Just as the Obama Administration was able to rescind many Bush Administration initiatives on energy and the environment, Obama's own environmental rules, executive agreements, and national monument declarations were vulnerable to repeal by Trump and the 115th Congress after the 2016 election. At best, these types of policy changes are partial and provisional.

The Pendulum Swing of the Environmental Presidency

Every recent president has acted in some way to expand the capacity of the office to influence environmental policy. Covering two decades of presidential involvement in environmental policy, this study explores the advances and retreats of the most recent administrations and analyzes the role of the presidency as it develops and becomes more active in shaping policy outcomes outside of the legislative process. The White House has been at the center of environmental policy ever since Richard Nixon established the Environmental Protection Agency, in an executive reorganization. The time of this agency consolidation and reorganization could also be characterized as an era of Congressional activism, when several major environmental statutes were passed, sometimes over veto threats of presidents. The environmental decade, which ended abruptly with the election of 1980, was followed by a decade of conflict that nevertheless managed to produce major environmental policies through the legislative process. Paradoxically, the Reagan assault on the regulatory state reinvigorated the environmental movement in opposition to that administration's policies, as the movement maintained pressure upon Congress to reauthorize and strengthen major environmental laws passed in the 1970s. This tense but productive decade culminated in the Clean Air Act Amendments of 1990, which established an innovative cap-and-trade program to control acid rain. Passed by a healthy bipartisan majority, the act was also the last major statute to require new command-and-control pollution standards.

The 1990s saw reauthorization of the 1974 Safe Drinking Water Act and the Food Protection Act, which repealed the controversial Delaney Clause, from the Pure Food and Drug Act. The latter was an urgent priority for the agriculture lobby and its resolution was a bipartisan cause in 1996, an election year. Subsequently, under Clinton and Bush, Congressional activism waned and an increasingly assertive executive branch shifted the balance of power between the branches of government. Ambitious bills to reform the nation's environmental laws and address new challenges have languished since the 1980s, whereas the most successful new policy initiatives to control pollution and manage natural resources have originated within the executive branch. Consequently, ENR policy is characterized by equilibrium, with neither side able to succeed in expanding, repealing, or reforming the existing statutory framework.[4] Within this equilibrium, there are swings that follow presidential elections, explored further below. Since Congress was no longer the agent that drove environmental policy after the 1980s, the focus of environmental advocacy became the bureaucracy as well as the White House itself. Baumgartner and Jones describe the policymaking system as punctuated by long periods of stability, although significant policy change is possible.[5] Under these circumstances, the gridlock that seems to paralyze the legislative arena fosters experimentation by the executive branch. As the following chapters will argue, policy direction is influenced by the leadership and agenda of federal bureaucracies, which in a sense are prizes for the winning electoral coalition and political party.

Much of the existing literature on presidential power and public policy is concerned with unilateral instruments, which include executive orders, proclamations, signing statements, and executive agreements. These are some of the policy instruments available to presidents, though of greater consequence for ENR policy are contextual tools, including: appointments, executive reorganizations, budgeting, and rulemaking. Taken together, they constitute the toolbox of the administrative presidency.[6] The strategic use of this multifaceted toolkit allows presidents to deliver on their promises in the absence of a supportive Congress or a mobilized public. The administrative strategy took on greater importance as a result of the deadlocked politics that emerged in the 1990s. Case studies of public land management, water pollution, toxic pollution, and climate change will illustrate in subsequent chapters the transformation that occurred as the nexus of ENR policymaking and innovation shifted from the legislative to the administrative arena.

After the "environmental decade" of the 1970s, the movement developed into a mature advocacy community by transforming from a grassroots movement seeking legislation to a set of entrenched, powerful interest groups organized to influence the administration through lobbying and strategic mobilization.[7] As an entrenched beltway interest, in the decades to follow the environmental community grew better positioned to interact with the executive branch, which asserted greater leadership over domestic policy by shaping implementation decisions as well as the agenda. Over time, gridlock made new legislation and creative reform of existing statutes nearly impossible. The movement itself seemed to diminish in scope and ambition, as its objective shrunk from broad societal transformation to a defensive posture, fighting against efforts on multiple fronts to dilute or repeal the nation's existing environmental laws.[8] The movement saw diminishing returns from mass membership over time, and the election of 2000 has been described as symbol of the movement's declining influence.[9]

Environmental groups became more specialized, professionalized, and accustomed to influencing policy at access points other than Congress. The federal courts, an important battleground since the movement's early years, became less hospitable to environmentalists' claims. Federal statutes such as the National

Environmental Policy Act (NEPA) and landmark Supreme Court decisions expanded the definition of standing and armed the advocacy community with new procedural instruments to challenge major projects. Environmental groups used these tools to enormous effect, but NEPA challenges declined in the 1980s, when judges became more deferential to government agencies as long as the procedural requirements of the law were met.[10] Furthermore, business levelled the playing field by forming their own public interest groups, such as the Western States Legal Foundation and the Scaife Foundation, to challenge environmental restrictions in the courts. Many judges were more sympathetic to business groups and property-rights advocates, and federal courts would frequently issue conflicting decisions. Rather than resolving longstanding policy disputes, the effect of environmental litigation was often to prolong conflict and establish the need for new rules, forcing the issues back into the administrative arena, as discussed in the case studies to follow.

Politicizing the Administration

The onset of legislative gridlock that renders domestic policy resistant to change also empowers the bureaucracy. Administrators have been able to use their discretion to influence policy according to their preferences and resist reform attempts in Congress. Consequently, the success of a major policy initiative requires the president to assert control over the administration.[11] Regulatory policies are especially prone to manipulation through the administrative process, as demonstrated below. Once Congress enacted policies addressing the environment, occupational health and safety in the 1970s, it was not long before a political stalemate developed.[12] The same dynamic affected public land management. As environmentalists, mining, and timber interests pressure Congress from both sides, battles over these policies in the legislative arena have been fought to a draw. This elevates the importance of such agencies as the U.S. Forest Service and Bureau of Land Management, and rulemaking has become the primary means of conflict management.[13]

Divided government is only part of the story, as every president since Clinton experienced setbacks with their environmental policy agendas despite beginning their terms in office with Congressional majorities of their own parties. For instance, as discussed further in Chapter 3, President Clinton shifted strategy after the 103rd Congress frustrated his effort to reform the nation's 120-year old mining law in 1994. Abandoning any hope of reform legislation in the Republican-controlled Congress, the Interior Department developed new rules to restrict hardrock mining, which took effect on Clinton's final day in office. Since the senate could not agree to raise royalties on minerals extracted from federal land, new rules were developed that required mine operators to seek approval before engaging in activities that would cause significant disturbance of the land and gave the Secretary of the Interior the authority to deny permits if operators failed to show that they would not cause undue environmental degradation.[14] The action reflected a view within the administration that lawmakers effectively ceded policymaking to the executive branch because of their failure to deliver badly-needed mining law reform through the legislative process. Noting that mining reform was hopelessly gridlocked for decades, Interior Department Solicitor John Leshy argued that administrative policymaking does not usurp Congressional authority when the legislative branch refuses to take action.[15]

Another factor in the onset of policy gridlock was the contentious nature of social regulation. Most federal regulatory policies were intended to maintain competitive markets until the 1970s, when Congress enacted many of the nation's major environmental statutes, as well as laws to protect health and safety. The most obvious difference between this form of regulation and traditional economic regulation is that social regulation is usually forced upon the industry at the request of public interest groups, whereas previously the competitive regulatory policies of the past were established at the request of the regulated industries, and were mainly concerned with prices and restricting entry into markets. There was less political conflict over the earlier generation of competitive regulatory policies, because the targets of those regulations generally agree on the goals, and policy is made in a stable political environment.[16] Independent agencies such as the EPA and the Office of Surface Mining were created to promulgate and enforce rules to implement newly-enacted environmental statutes. Their rules fall under the category of social regulation because they tend to have an adversarial relationship with the entities they regulate. By the end of the environmental decade of the 1970s, business groups had shifted their strategy from containing the regulatory state to a full-scale counterattack, expressed in the election victory of Ronald Reagan.[17] Subsequently, business groups pressured Congress to rein in regulatory agencies, and battles over specific regulations played out in the courts.

Of course, the rise of social regulation in the 1960s–1970s produced a backlash that altered the political environment in the 1980s. Agencies such as the EPA and the FTC that flexed their regulatory muscles were often met with resistance from lawmakers.[18] Congress was in an antiregulatory mood by 1980, when the House of Representatives debated appropriations cuts and a legislative veto to rein in the agency. Because these protective regulatory policies continue to be popular with the public, Congressional opponents have been unsuccessful in their efforts to rescind them. Predictably, the result is a long-term deadlock in which neither side has been able to reform or repeal these venerable statutes.[19]

Legislative policy change is unlikely to occur in the absence of a mobilized public, a favorable political climate, a supportive and powerful policy subsystem, and a commonly-accepted narrative with clearly-defined policy problems, causes, and solutions.[20] Again, somewhat paradoxically the bureaucracy has been empowered by the inability of a deadlocked Congress to act decisively. Because many regulatory policies give federal agencies significant responsibilities for implementation, administrators enjoy considerable discretion to determine the direction of those policies.[21] Presidents thus have a powerful motivation to assert control over the bureaucracy in order to ensure their policy initiatives succeed.

Following the rise of the Administrative Presidency, recent presidents are more likely to avoid the traditional legislative route for their environmental policy agendas in favor of a strategy that relies on unilateral and contextual tools. Not only is the administrative strategy a low-risk alternative to a bruising battle in the legislative arena, the mere threat of unilateral action can pressure Congress to act on the president's priorities. Republican administrations since the 1980s have more often used administrative strategies to scale back ENR policies, while Democratic administrations tend to use them to further the goals of conservation and pollution reduction. At the same time, gridlock has diminished the once-dominant role of Congress. Overall, the pattern is one of increasing presidential dominance and policymaking through administrative discretion.

One consequence of persistent stalemates over environmental legislation is more assertive administrative policymaking. As the politics of environmental regulation have grown more contentious, presidents and their appointees have less reason to become entangled with debates over legislation. Over the past four decades, it has become safer politically for presidents to wait and try to influence these policies at the implementation stage. Even though the executive branch is granted a limited policy role under the Constitution, presidents have expanded their policy leadership role by exercising their influence over the bureaucracy. The concentration of executive control over environmental policymaking has continued to develop, even evolve. In the Obama Administration, authority was delegated to White House advisors not subject to Senate confirmation "czars" to manage presidential priorities such as energy, the environment, and green jobs. As the examples of Bill Clinton and Barack Obama explored below demonstrate, administrative strategies may be used to pursue a pro-environment agenda in the face of legislative gridlock and obstruction. Conversely, presidents have also used administrative tools to advance an anti-regulatory agenda by constraining the bureaucracy.[22] The Trump Administration has continued this trend, relying upon informal "deregulation teams" and unconfirmed "acting" administrators to implement its agenda with little Congressional oversight.

Why Study Presidential Administration?

In arguing that the "power to persuade" was the essence of presidential leadership, Richard Neustadt voiced a prevailing view that the president's influence over policymaking stemmed from his role as legislative leader. Limited to a narrow set of Constitutional powers and constrained by checks and balances, the office was considered too weak to enact a policy agenda without enlisting the support of Congress, interest groups and the judiciary.[23] There are also a number of tools available to the presidency that can be used to make policy independently. Some of these are powers of the office and others are strategies developed by presidents themselves. Contemporary presidents have often turned to these instruments as an alternative to bargaining. They could have any number of motivations for bypassing Congress, including the desire to act quickly and decisively, avoiding the risk of their legislative proposals becoming bogged down in debate, or the need to act in the face of Congressional opposition. For these purposes, presidents have a range of prerogative powers.[24] In the policy areas where Congress has most clearly delegated authority, presidents are more apt to use executive orders as the instrument of choice.[25] Agency rulemaking is used to further presidential priorities across most policy areas. As Chapter 2 explains, an elaborate structure has been cultivated within the Executive Office of the President to review regulations and manage the regulatory process.

Due to their resources and expertise, bureaucracies have been delegated significant authority over the policies they administer. Environmental agencies, in particular, are in a strong position to influence policy outcomes because of the large number of responsibilities under federal law, as well as the discretion granted by their enabling statutes and the Administrative Procedures Act. Regulations are the product of an assembly process and are more likely to reflect the expertise of the regulators. In addition to the standard implementation responsibilities of standard setting, monitoring and enforcement, the bureaucracy has become a significant actor in the policy process. Because of the growth in the number and scope of federal programs after the 1960s, the bureaucracy has gradually expanded its policymaking role. Consequently, more policymaking occurs within the permanent government, without the accountability or opportunities for democratic deliberation characteristic of the legislative process.

Because environmental agencies, particularly the EPA, were designed to be independent and to resist political pressures, most of their decisions have been resilient enough to withstand endless legal and political challenges. The relative independence of the EPA, while giving its scientists and regulators further credibility in the face constant criticism, also makes it resistant to direct control by either the legislative or executive branches. This illustrates a central problem of governance: the success of any presidential initiative depends upon the acceptance and cooperation of the government agencies that share responsibility for implementation of the policy, as well as a host of competing mandates. Presidents must use the resources of the White House to press for their policy preferences in order to prevent their decisions from becoming stalled in the vast machinery of the bureaucracy or diluted by other priorities.[26] Thus, there is a strong incentive to politicize the bureaucracy. To the extent that agency leaders share the president's priorities, their agencies are more likely to march to a single drummer. In a challenge to the Administrative Presidency thesis, David Rosenbloom argues that agencies are largely insulated from direct White House control. From this perspective, Congress is the stronger principal because of oversight, appropriations and the delegation of authority from the enabling statute.[27] Indeed, rulemaking involves the exercise of a legislative power delegated by Congress, and this consideration should outweigh the influence of presidential appointees, who may or may not be able to impose a policy agenda that diverges from the agency's legislative mandates. Although candidates are not always chosen for patronage positions on the basis of their loyalty to the president's policy agenda, the examples of Trump appointees and Bush/Cheney appointees who pressed their energy development agenda shows top administration personnel can be effective in advancing top White House priorities, as discussed in Chapter 2.

The environmental administrative presidency has also been a source of innovation. Despite the political stalemate in the legislative branch, presidents have had some success encouraging alternative policy approaches within the bureaucracy through their appointment and budgeting powers and rulemaking. Presidents Clinton, Bush, and Obama, as well as their appointees, encouraged the use of alternative policy instruments to manage conflicts and reduce pollution without new command-and-control legislation. The 1990s saw an expansion of customized policies, especially negotiated rulemaking, performance standards and alternative standards.[28] Other next-generation approaches include market-based policies and information technology. Most recently, the executive bureaucracy cultivated these new approaches and applied them to new and emerging environmental problems.

The mostly command-and-control policies of the 1970s emphasizing compliance with standards had run their course, and continued progress is only possible if new approaches are adopted.[29] Persistent gridlock prevented legislative solutions and the consensus for new environmental legislation has not materialized, despite some speculation that politics in the 21st century would de-emphasize conflict and usher in a new generation of collaborative policies.[30] Instead, policy whiplash has become the norm. Regulatory agencies often take the initiative but are hastened and slowed by gridlocked politics. Caught between competing principals, the EPA and natural resource agencies have had limited success introducing innovative environmental policies. Consequently, most of the recent examples of innovation and regulatory reinvention, starting with the EPA's Project XL in 1995 and beyond, have been the result of a coordinated administration under presidential leadership.

Green Drift and Retrenchment

The interests seeking to undermine those environmental programs have also looked to the executive branch. Because the policies made with the stroke of one president's pen may be altered or undone by another, environmental policy can meander back and forth. Klyza and Sousa argue that despite all these halts and starts, the larger current has been moving in an increasingly green direction.[31] Conservative administrations and Congresses brought about waves of deregulation, but none have succeeded in repealing or significantly weakening the major environmental statutes of the 1960s and 1970s. Moreover, presidents and agencies have used administrative tools to strengthen those programs and extend their protections to new populations and to address new and emerging environmental threats. This phenomenon, which Klyza and Sousa call "green drift," is most likely to occur when statutes have clear mandates, deadlines, and enforcement provisions, and especially when presidential administrations support the statutes' goals.[32]

The alternative pathways used to circumvent gridlock have also created opportunities for policy retrenchment. Judith Layzer argued the favored strategies for retrenchment involve reining in the administrative state through procedural requirements, including cost-benefit analysis and regulatory review, as well as using the rulemaking process to change the substance of regulations to weaken their impact or limit their scope.[33] The statutory framework remains intact, but business interests and their allies in the White House have succeeded in manipulating the tools of the administrative presidency to undercut those policies at the implementation stage. Furthermore, these quiet victories were fought and won not in the halls of Congress, but in venues such as the federal courts, the White House Office of Information and Regulatory Affairs (OIRA) and the pages of the Federal Register.

The Politics of Ambivalence

The first Earth Day in 1970 sent a strong message that national leadership was needed to address the environmental crisis. The modern environmental movement organized in response to a series of ecological catastrophes and severe pollution events, in a time when political protest was more common and more effective. Consequently, President Nixon and Congress responded with several major environmental statutes that preempted state laws, including the Clean Air Act and National Environmental Policy Act (1970), and the Clean Water Act (1972). Politicians took credit for their achievements protecting the environment, but in many cases the victories were merely symbolic. Public concern for the environment dissipated after Congress enacted the first major pollution laws. Even as federal agencies labored to write and enforce regulations, many of these policies were weakened and the goals proved difficult to enforce or were not met.

Public opinion continued to be in favor of environmental protection, but the intensity of this softened as other problems displaced the environmental crisis. Figure 1.1 shows the proportion of respondents that prioritized environmental protection over economic growth. Between 1991 and 2011, the proportion that prioritized environmental protection declined from 71 percent to 36 percent. For the first time in 2008, a higher proportion of respondents prioritized economic growth.

Environmental leadership can be politically risky for elected officials, even when they pursue popular initiatives. Americans simply do not "vote green," as environmental issues seldom register among the most important problems in national elections.[34] Nevertheless, there is some evidence that the environment can be used effectively as a wedge issue to attract centrist voters in the context of

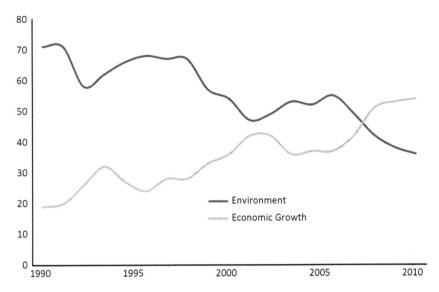

FIGURE 1.1 Percentage of Respondents Prioritizing Environmental Protection over Economic Growth, 1990–2011

certain campaigns. This voter response happened in 1996, when Clinton used his executive authority to enact a small number of conspicuous conservation policies as he campaigned for re-election. Most of Clinton's environmental agenda had stalled early in his first term, and neither he nor Republican candidate Bob Dole had been seen as strong environmentalists prior to the enactment of these greener policies. During his campaign, Clinton took credit for a sharp increase in the cleanup of federal Superfund sites and the redevelopment of urban brownfields. He also made a highly publicized election-year national monument designation to shore up his support among environmentalists.

Actions such as those Clinton took during his campaign for re-election represented more than just political maneuvering. They were opportunities to move forward with the implementation of policy goals when progress was otherwise blocked by deadlocked politics. In one case, presidential leadership removed obstacles and accelerated the cleanup of several polluted sites in urban areas, despite some local concerns about weakened pollution standards. In the other, the president took sides in a longstanding land preservation dispute, over the objections of state and local governments, after Congress and the courts had been unable to settle it. As the case studies analyzed in the following chapters show, the effects of these initiatives have endured beyond the immediate controversies because they were popular with important constituencies and difficult to overturn without an act of Congress.

Environmental Advocacy

As the traditional, already somewhat rocky pathways for environmental policy-making became further blocked with gridlock, the advocacy community adapted with new strategies and tactics. Groups became more sophisticated and professionalized, more able to breach the impasse of gridlock in response to the changes that swept the federal government following the 1980 election.[35] Soon after Reagan's inauguration, the Chief Executive Officers of the ten largest environmental organizations began to hold regular meetings in Washington, DC, where they formulated their response to the Reagan agenda. The "Group of Ten" met the challenge by growing in size and professionalism, and coordinating their activities.[36] For these advocacy groups, coalition building became a powerful tool against gridlock.

The process of institutionalizing the movement echoed the actions of the business lobby during the previous decade. It resembled the Business Roundtable, a forum that united disparate business interests under a single umbrella, channeling their resources for the counterattack upon the regulatory state that culminated in the triumph of Reagan's election. But this time, the Group of Ten also served a similar purpose for environmentalists. The institutionalization process allowed leaders of the movement to redefine their roles from outsiders to executives who were defenders of a system.[37] Table 1.1 lists the largest national environmental organizations and the size of their professional staffs. New environmental organizations were transformed, with large professional staffs and multimillion-dollar budgets. These groups were forced to smooth over their differences and create a common mainstream identity. As Robert Gottlieb wrote, "the Ten, in existence for less than a decade, had effectively redefined mainstream environmentalism less as a movement and more directly as an adjunct to the policy process."[38] After expanding in the 1980s, established organizations began to resemble the newer ones. In contrast to the movement that challenged the status quo during environmental decade, through coalition building and professionalization, environmental groups became essential players in the issue networks that develop and maintain policy.

Regulatory activity now accounts for a significant share of the policy and political agenda of administrations.[39] This has not escaped the attention of interest groups, which are more likely to identify administrative agencies as prime targets for lobbying efforts. By submitting written comments on proposed rules, participating in regulatory negotiations and lobbying agencies directly, interest groups affect policy by influencing the rulemaking process. Surveys of business groups have found that most consider this to be an effective method of influencing agency decisions.[40]

Lobbying the executive branch makes as much sense from the perspective of environmentalists. Despite their increased professionalization, the major environmental groups continue to enjoy strength in numbers. These groups can mobilize their networks of supporters effectively to create a groundswell of support for (or opposition to) legislative proposals, and key representatives can feel pressure if those

TABLE 1.1 Professional Staff of Environmental Groups, 1970–2011

	1970	1980	1990	2000	2011
Sierra Club	60	145	250	290	500
NRDC	-	77	128	195	400
EDF	-	41	110	230	324
WWF (US)	-	-	244	392	439
Nat. Wildlife Fed.	-	464	800	692	400
Audubon Society	275	n/a	315	550	520
NPCA	-	-	43	89	157
Isaak Walton League	-	-	23	38	35
Wilderness Society	-	-	135	118	146
Ducks Unlimited	-	-	275	622	450
Defenders of Wildlife	-	-	34	121	101
Nature Conservancy	-	77	1,150	2,773	2,600
LCV	-	5	67	25	n/a
Greenpeace USA	-	-	250	90	70
Friends of the Earth	12	70	45	32	33
Earthjustice	-	10	n/a	113	143
American Rivers	-	-	38	53	64
Clean Water Action	-	-	90	90	450
Rainforest Action Net.	-	-	10	38	43
Ocean Conservancy	-	-	38	63	42
Conservation Int.	-	-	57	534	1,250
Conservation Fund	-	-	-	75	79
Trust for Public Land	-	-	-	310	400

supporters are their constituents. This numerical advantage helps environmental groups compete for influence with business groups, which enjoy the advantage of deep pockets. From 2000 to 2016, corporations spent more than $2 billion lobbying Congress over climate legislation.[41] Much in the way environmental organizations use their ranks to influence legislative votes, they can also summon the energy of their mass membership in a show of public support to influence policy at the implementation stage, where the executive branch is most active.

Moreover, because the executive branch is active at every stage of the policy process groups have more opportunities to influence policy outcomes by focusing their energy on the White House and the bureaucracy. For environmental advocates both inside and outside government, presidential support for a proposal is coveted because of the president's powers to set the agenda, to propose and veto legislation, and to issue directives to the federal bureaucracy. Furthermore, it pays to court the White House because of its potential to influence implementation decisions. Table 1.2 summarizes the largest national environmental groups' efforts to mobilize their members in 2012. In most of these "actions," group members were notified by e-mail and urged to contact a government official about a particular policy issue, or to take some other action en masse.

Even though Congress was most often the focus, a significant number of actions directed members to contact the Obama Administration. Of the 1,834 member actions, 21 percent were directed toward the bureaucracy and another 18 percent toward the White House. These combined for 39 percent of member actions in 2012, compared to 36 percent that directed members to contact their representatives in Congress. Larger keystone organizations urged more actions overall.

Most environmental organizations were just as likely to focus on the White House or the bureaucracy. Only a few established sectoral players were more likely to focus on Congress, including the Audubon Society, the Environmental

TABLE 1.2 Environmental Group Member Actions, 2012

	Contact	Contact	Contact	Contact State	Consumer Actions	Other Actions
	Congress	Agency	President			
Sierra Club	156	96	100	20	3	1
NRDC	41	31	16	5	2	0
EDF	59	34	26	0	8	3
WWF	50	12	18	0	57	5
Nat. Wildlife Fed.	9	7	10	0	0	2
Audubon Society	43	10	1	0	0	0
NPCA	35	23	22	0	6	2
Isaak Walton League	27	1	0	0	0	0
Wilderness Society	0	6	3	0	0	1
Ducks Unlimited	22	0	0	0	0	0
Defenders of Wildlife	13	30	5	10	0	0
Nature Conservancy	12	0	0	0	0	0
LCV	36	9	7	0	5	4
Earthjustice	84	46	58	40	7	3
American Rivers	37	50	22	31	0	0
Clean Water Action	43	14	15	81	4	3
Ocean Conservancy	18	13	4	1	0	0
Rainforest Action	0	10	18	6	73	5
Greenpeace USA	0	3	7	0	3	1
Friends of the Earth	8	14	8	0	0	0
Total	693	409	340	194	168	30
	(36%)	(21%)	(18%)	(14%)	(9%)	(2%)

Defense Fund, the Isaak Walton League, as well as the Nature Conservancy, the Ocean Conservancy and Clean Water Action. Another exception is the League of Conservation Voters, whose primary mission is involvement in Congressional elections. Smaller niche players had fewer actions overall (with the exceptions of Earthjustice and American Rivers) and were more likely to direct action toward state governments. Only a small percentage of actions involved activities other than contacting government officials. Most of these exceptions were consumer actions, such as boycotts, or instructions to contact business leaders to express thanks or admonishment for corporate behavior. Although consumer actions were employed by several groups, the majority were urged by smaller groups whose niche was promoting change by influencing consumer behavior.[42]

A similar pattern of interest group activity can be observed in a non-presidential election year. Table 1.3 shows that the big national environmental

TABLE 1.3 Environmental Group Member Actions, 2018

	Contact	Contact	Contact	Contact State	Consumer Actions	Other Actions
	Congress	Agency	President			
Sierra Club	24	29	1	50	9	17
NRDC	8	12	5	2	0	1
EDF	5	7	0	0	0	0
WWF	5	3	2	0	1	2
Nat. Wildlife Fed.	15	4	1	2	0	0
Audubon Society	12	1	0	1	0	0
NPCA	9	10	1	3	0	0
Isaak Walton League	8	0	0	0	0	0
Wilderness Society	0	2	0	0	0	0
Ducks Unlimited	2	0	0	0	0	0
Defenders of Wildlife	10	5	0	1	0	0
Nature Conservancy	9	5	5	0	0	0
LCV	8	6	0	1	1	0
Earthjustice	16	20	2	2	0	0
American Rivers	8	4	0	3	0	2
Clean Water Action	9	5	0	0	3	1
Ocean Conservancy	2	2	0	0	0	0
Rainforest Action	0	0	0	0	0	0
Greenpeace USA	0	1	0	1	9	1
Friends of the Earth	10	1	0	0	2	0
Total	160	117	26	84	25	24
	(37%)	(27%)	(6%)	(19%)	(6%)	(5%)

groups had fewer member actions overall during the midterm election year of 2018. Of the 436 member actions, 37 percent were directed at Congress, and 27 percent at federal agencies. The major difference between that year and 2012 is that only 6 percent of the actions urged the members to contact the president; almost certainly because of the relative efficacy of lobbying Trump compared to Obama. In 2018, there were also far more actions directed at state governments (19 percent) which were likely seen as more fertile ground for environmental progress. Several of these state-level actions were indirectly focused on national policy, as environmental groups encouraged members to support lawsuits against the Trump Administration by California and New York, which had become bulwarks of resistance.

Plan of the Book

The following chapter considers the expanding role of presidential administration in ENR policy. All four recent presidents have been assertive in the use of administrative tools to advance their environmental priorities. Clinton, Bush, Obama, and Trump have employed administrative authority and strategies tested by their predecessors, often to greater effect. These instruments will be reviewed, along with unilateral powers, and various reforms that recent presidents have used to expand the influence of the office over ENR policy. Each has put his own mark on the institutional presidency in the process.

The subsequent chapters present case studies of selected areas of ENR policy, showing how Congressional inertia has given way to presidential initiative. The four cases represent a cross-section of environmental problems, ranging from natural resources to various forms of pollution affecting the environment through different media. The cases also represent a mixture of old and new challenges. They vary from the long-simmering conflicts over the outdated statutory framework for public land management, to the 1970s-era policies to control water pollution, to the more current conflicts over toxic contamination, to the emerging threat to the global climate. Chapter 3 examines the role of presidential administration in public land management. Following an era when a conservationist agenda fueled major legislation, such as the Wilderness Act, NFMA and FLPMA, Congress turned antiregulatory. During the Clinton years, the green state became the defender of conservation policies. Clinton used a signing statement to blunt the impact of a legislative assault on the Endangered Species Act, and he made broad use of presidential directives and executive orders for land preservation. Clinton also championed proposals by career administrators at the BLM and Forest Service to restrict hardrock mining and preserve roadless forests.[43] Even George W. Bush, a president whose background was in management, was more comfortable using administrative tools to promote a pro-development agenda. Bush used his administrative strategy to pursue conservation goals as well, notably protecting the Northwest Hawaiian Islands with the 2006 proclamation of a national monument.

Water quality, the focus of Chapter 4, is one area in which administrative leadership has encouraged new policy approaches. The discharge permit system established under the Clean Water Act succeeded in curtailing the most visible surface water pollution from industrial and municipal sources, but it was ineffective against the growing problem of nonpoint sources. Following the Clinton Administration's reinvention initiative, the EPA began to experiment with negotiated rulemaking and unified rulemaking to set standards for several drinking water contaminants. Later, the EPA began to address the growing problem of stormwater runoff. Here, Congress had offered little direction, but the EPA developed regulations under the authority of the CWA Amendments of 1987. Chapter 4 also reviews how the rule requiring permits for concentrated animal feeding operations was promulgated at the end of the Clinton years, then withdrawn and scaled back by the Bush Administration, and revised again by the Obama Administration. Finally, in 2011, the EPA proposed a permit system for runoff under the EPA's existing statutory authority, which was later suspended by the Trump Administration.

Focusing then on the control of toxic and hazardous substances in the face of legislative gridlock, Chapter 5 examines how the green state took the initiative and adopted new regulatory approaches. New partnerships were formed during the Clinton years, leading to successful state-level recycling and brownfield reclamation programs. Even though George W. Bush had the benefit of a Republican Congress for most of his presidency, he avoided new environmental legislation, preferring to use unilateral actions to pursue his agenda. The Bush Administration stressed voluntary policies to control hazardous waste and used rulemaking to relax standards for hazardous air pollutants, going so far as (unsuccessfully) proposing a cap-and-trade market for mercury emissions. In recent years, regulation of hazardous substances benefited from the widespread integration of information technology. One example analyzed in this chapter is the Toxic Release Inventory (TRI). Created by an act of Congress in the 1980s, the TRI was considered one of the most cost-effective national environmental programs, and making the database available online has made it more accessible to the public and advocacy groups.[44]

Chapter 6 discusses the critical role of presidential leadership on climate change. Ever since the 1990s, influential members of Congress have supported legislation to create a cap-and-trade market to curb carbon dioxide emissions as an alternative to the command-and-control regime available under the Clean Air Act. However, those efforts stalled because of objection from business groups, concerns about cost, and Congressional gridlock. With the legislative debate going nowhere, climate change became a matter of presidential politics in the 2000 election. Upon becoming president, George W. Bush quickly yielded to the same forces, abandoned a campaign pledge to limit curb carbon emissions, and withdrew from the Kyoto Accord. The White House then actively discouraged the development of greenhouse gas policy and suppressed information about climate change for eight years. The initiative for climate change policy remained with the executive branch following the election of Barack Obama, who pivoted to an administrative strategy after Congress failed to enact cap-and-trade legislation for GHG emissions. The EPA sought to regulate carbon emissions for the first time under Obama's Clean Power Plan, until it was suspended by the Trump Administration in favor of a voluntary reduction plan.

The concluding chapter reviews the case studies and explains how the environmental policy issue network has evolved over time. The review and analyses of the cases will allow a clearer assessment of the development of the institution of the presidency and its expanding influence within the network. The final chapter will also include an early assessment of President Trump and projections based on his first three years. Based on his administrative actions, Trump's approach to environmental policy most resembles George W. Bush's. After avoiding environmental issues in the campaign, he quickly moved on an antiregulatory agenda upon assuming office. Looking forward, the study will suggest that taking the focus off the environment makes it more likely that Trump will attempt to deliver upon his promises quietly through unilateral action, much the way his predecessors did, instead of spending precious political capital with Congress. Finally, the study examines the complex policy implications of the new political landscape for policymakers and activists alike.

Conclusion

Due to a countermobilization of interests after the environmental decade of the 1970s, the movement no longer wields the same influence. The resulting political stalemate has reduced the potential for victories in Congress and the courts, making the executive branch the most promising arena to press for policy change. In 1990, the Clean Air Act was amended to expand the scope of federal air pollution controls, mandating standards for over nearly two hundred hazardous air pollutants and addressing the problems of acid rain and stratospheric ozone. For more than two decades since that landmark statute, Congress has been reluctant to reclaim the leadership role that it once claimed. The heyday of grassroots activism was a distant memory, leading some observers to proclaim "the death of environmentalism."[45]

The transformation of the political landscape is apparent in the failures of recent presidents to win legislation. For environmentalists, President Obama's first term was about as disappointing as President Clinton's. After taking office with an ambitious environmental agenda, both became preoccupied with other priorities, namely the economy. Despite large majorities in both houses of Congress, both suffered defeats when major environmental legislation stalled. After those early defeats, both were more successful when they pursued their environmental priorities through administrative channels. Likewise, President George W. Bush relied on administrative rulemaking to relax air pollution regulations on power plants and increase energy and timber production on federal land. He, too, was frustrated in his attempts to move major environmental legislation through Congress, despite Republican majorities during most of his presidency. The Trump presidency also pursued an agenda of policy retrenchment, but unlike Bush, Trump had little interest in new legislation to carry it out.

The executive branch has most often presented a venue for environmental policy change when other pathways were closed off. The Administrative Presidency provided a foundation that recent presidents have built upon, and each one in turn expanded the capacity of the office to make environmental policy. Bill Clinton used executive orders to form new advisory councils on environmental justice and sustainable development, launched a series of administrative reforms that centralized control over environmental policy implementation, and introduced new voluntary programs and partnerships to manage resources and reduce pollution. George W. Bush, who prioritized energy and natural resource development over environmental protection, further centralizing his administration in a concerted effort to remove regulatory barriers. Bush involved Congress as little as possible, developing new strategies to affect policy change, and borrowing alternative policy models from the Clinton Administration when they suited his goals. As the next chapter will argue, both Barack Obama and Donald Trump were beneficiaries of this institutional development. Faced with many of the same obstacles and political constraints as their predecessors, Obama made full use of existing arrangements and strategies to pursue a broad range of environmental policy goals, and Trump was able to turn those same institutional arrangements into a multifaceted toolkit for policy retrenchment.

Notes

1 Donald Trump, "State of the Union Address", January 30, 2018, www.presidency.ucsb.edu/documents/address-before-joint-session-the-congress-the-state-the-union-25.
2 "The Power of the President: Recommendations to Advance Progressive Change" (Forward by John D. Podesta), Center for American Progress, November 15, 2010.
3 Obama had already endorsed the solar power idea during a visit to a 140-acre solar array at Nellis AFB in Nevada. By the fall of 2010, the Air Force only had plans for such facilities at three other bases. See "Air Force to Quadruple Solar Energy Production" (October 15, 2010). Inside AF: www.af.mil/News/Article-Display/Article/115298/air-force-to-quadruple-solar-energy-production/.
4 See Robert Repetto, ed., 2006, *Punctuated Equilibrium and the Dynamics of U.S. Environmental Policy*, New Haven, CT: Yale University Press.
5 Frank Baumgartner and Bryan Jones, 1993, *Agendas and Instability in American Politics*, Chicago: University of Chicago Press.
6 Richard P. Nathan, 1983, *The Administrative Presidency*, New York: Wiley.
7 Christopher Bosso, 2005, *Environment Inc.*, Lawrence, KS: University of Kansas Press.
8 Michael Shellenberger and Ted Nordhaus, 2004, "The Death of Environmentalism: Global Warming Politics in a Post-Environmental World," www.thebreakthrough.org/articles/the_death_of_environmentalism.
9 Phillip Shabecoff, 2001, *Earth Rising: American Environmentalism in the 21st Century*, Washington, DC: Island Press.

10 Walter A. Rosenbaum, 2012. *Environmental Politics and Policy*, 8th edition, Washington, DC: Congressional Quarterly Press.

11 Jeffrey Berry and Kent E. Portney, 1995, "Centralizing Regulatory Control and Interest Group Access: The Quayle Council on Competitiveness," in *Interest Group Politics*, 4th edition, edited by Allan J. Cigler and Burdett A. Loomis, Washington, DC: Congressional Quarterly Press.

12 These are among the policy areas most likely to be delegated by Congress. Epstein and O'Halloran show that in addition to appropriations committees, Congressional committees on environment, energy and labor are the most likely to hold oversight hearings. See David Epstein and Sharyn O'Halloran, 1999, *Delegating Powers*, Cambridge: Cambridge University Press.

13 Christopher McGrory Klyza and David Sousa, 2008, *American Environmental Policy, 1990–2006*, Cambridge, MA: MIT Press.

14 According to Leshy (2001) it was the administration's intent to bypass a gridlocked Congress. Once George W. Bush became president, the hardrock mining rule was reversed by an act of Congress. See John Leshy, 2001, "The Babbitt Legacy at the Department of the Interior: A Preliminary View," *Environmental Law* 31(2): 199–228.

15 Ibid.

16 Edward P. Fuchs, 1988, *Presidents, Management and Regulation*, Englewood Cliffs, NJ: Prentice Hall.

17 Andrew Szasz, 1984, "Industrial Resistance to Occupational Safety and Health Legislation: 1971–1981," *Social Problems* 32(2):103–16.

18 For example, when the Federal Trade Commission began issuing rules in the 1960s, it followed decades of restraint in which the FTC exercised its power to resolve disputes only on a case-by-case basis. The most notable early example was a 1964 rule requiring warning labels on cigarette packages. An immediate backlash from the tobacco industry resulted in an act of Congress that nullified the regulation. The FTC became more assertive about issuing regulations to protect consumers only after it was granted an explicit authority to do so under the Magnusson-Moss Warranty Act in 1974, but this did little to insulate the bureaucracy from criticism that the FTC exceeded its statutory authority. See William F. West, 1982, "The Politics of Administrative Rulemaking," *Public Administration Review* 42(5): 420–426.

19 Deadlocked policy areas are also the most likely to be delegated by Congress. Epstein and O'Halloran show that in addition to appropriations committees, Congressional committees on environment, energy and labor are the most likely to hold oversight hearings. See Epstein, David and Sharyn O'Halloran 1999. *Delegating Powers.* Cambridge: Cambridge University Press.

20 John Kingdon, 1995, *Agendas, Alternatives and Public Policies*, 2nd edition, Harper Collins.

21 Jeffrey Berry and Kent E. Portney, 1995, "Centralizing Regulatory Control and Interest Group Access: The Quayle Council on Competitiveness," in *Interest Group Politics*, 4th edition, edited by Allan J. Cigler and Burdett A. Loomis, Washington, DC: Congressional Quarterly Press.

22 Various studies of the administrative presidency have examined the presidencies of Nixon: Richard Nathan's *The Plot That Failed*, Robert Durant's study of Reagan and the BLM, *The Administrative Presidency Revisted,* Charles Tiefer's study of George H. Bush, *The Semi-Sovereign President,* and Elena Kagan's article on the Clinton Administration. See Richard P. Nathan, 1975, *The plot that failed*, Wiley. See Robert F. Durant, 1992, *The Administrative Presidency Revisited: Public Lands, the BLM, and the Reagan Revolution*, Albany: SUNY Press. See Charles Tiefer, 1994, *The semi-sovereign presidency: the Bush administration's strategy for governing without Congress*, Boulder: Westview Press. See Elena Kagan, 2001, "Presidential Administration," *Harvard Law Review* 114: 2245–2385.

23 For the now dated but classic study/seminal work on the power of persuasion, see Richard E. Neustadt, 1991, *Presidential Power and the Modern Presidents: The Politics of Leadership from Roosevelt to Reagan,* 4th edition, New York: Free Press.

24 See William G. Howell, 2003, *Power Without Persuasion: The Politics of Direct Presidential Action*, Princeton, NJ: Princeton University Press. Also see Phillip J. Cooper, 2002, *By Order of the President*, Lawrence, KS: University Press of Kansas, and Steven Shull, 2006, *Policy by Other Means*, College Station, TX: Texas A&M University Press.

25 See Kenneth Mayer, 2001, *With the Stroke of a Pen*, Princeton: Princeton University Press. See Adam Warber, 2006, *Executive Orders and the Modern Presidency: Legislating from the Oval Office*, Boulder, CO: Lynne Rienner Publishers.

26 Terry M. Moe, 1985, "The Politicized Presidency" in *New Directions in American Politics*, edited by John E. Chubb and Paul E. Peterson, Washington, DC: Brookings.

27 David H. Rosenbloom, 2000, *Building a Legislative-Centered Public Administration*, Tuscaloosa, AL: University of Alabama Press.

28 Daniel J. Fiorino, 2006, *The New Environmental Regulation*, Cambridge, MA: MIT Press.

29 Ibid.

30 For example, see Mary Graham, 1999, *The Morning After Earth Day*, Washington, DC: Brookings.

31 Christopher McGrory Klyza and David Sousa, 2008, *American Environmental Policy, 1990–2006*, Cambridge, MA: MIT Press.

32 Ibid.

33 Judith. A. Layzer, 2012, *Open for Business: Conservatives' Opposition to Environmental Regulation*, Cambridge, MA: MIT Press.

34 For a discussion of the weak impact of environmental attitudes on presidential elections, see Chapter 6 of Deborah L. Guber, 2003, *The Grassroots of a Green Revolution*, Cambridge, MA: MIT Press.

35 Michael E. Kraft, 2000, "U.S. Environmental Policy and Politics: From the 1960s to the 1990s," *Journal of Policy History* 12(1): 17–42.

36 The "Group of Ten" was a coalition of established conservation groups, the National Wildlife Federation, Isaak Walton League, Audubon Society, Sierra Club and the Wilderness Society, as well as five newer groups: the Natural Resources Defense Council, Environmental Defense Fund, Environmental Policy Center, Friends of the Earth, and the National Parks Conservation Association. See Robert Gottlieb, 1993, *Forcing the Spring*, Washington, DC: Island Press.

37 Ibid.

38 Ibid.

39 Elena Kagan, 2001, "Presidential Administration," *Harvard Law Review* 114: 2245–2385.

40 Studies by Furlong and Kerwin have established that interest groups can be just as active in their efforts to influence regulatory agencies as they are in lobbying legislators. See Scott R. Furlong, 1997, "Interest Group Influence on Rule Making," *Administration and Society* 29(3): 325–47. See Cornelius M. Kerwin, 2003, *Rulemaking: How Government Agencies Write Law and Make Policy*, 3rd edition, Washington, DC: Congressional Quarterly Press.

41 Robert J. Brulle, 2018, "The climate lobby: a sectoral analysis of lobbying spending on climate change in the USA, 2000 to 2016," *Climactic Change* 149: 289–303.

42 The Rainforest Action Network was especially active in this area 2012, but it ceased this type of action by 2018. Therefore it is not listed in Table 1.3.

43 See Michael P. Dombeck, Christopher A. Wood and Jack E. Williams, 2003, *From Conquest to Conservation*, Washington, DC: Island Press. See Tom Turner, 2009, *Roadless Rules: The Struggle for the Last Wild Forests*, Washington, DC: Island Press.

44 See Michael E. Kraft, Mark Stephan and Troy Abel, 2011, *Coming Clean*, Cambridge: MIT Press.

45 Michael Shellenberger and Ted Nordhaus, 2004, "The Death of Environmentalism: Global Warming Politics in a Post-Environmental World," www.thebreakthrough.org/articles/the_death_of_environmentalism.

Bibliography

"Air Force to Quadruple Solar Energy Production" (October 15, 2010). Inside AF: www.af.mil/News/Article-Display/Article/115298/air-force-to-quadruple-solar-energy-production/.

Baumgartner, Frank and Bryan Jones. 1993. *Agendas and Instability in American Politics.* Chicago: University of Chicago Press.

Berry, Jeffrey and Kent E. Portney. 1995. "Centralizing Regulatory Control and Interest Group Access: The Quayle Council on Competitiveness." in *Interest Group Politics*, 4th edition, edited by Allan J. Cigler and Burdett A. Loomis. Washington, DC: Congressional Quarterly Press.

Bosso, Christopher. 2005. *Environment Inc.* Lawrence, KS: University of Kansas Press.

Brulle, Robert J. 2018. "The climate lobby: a sectoral analysis of lobbying spending on climate change in the USA, 2000 to 2016." *Climactic Change* 149: 289–303.

Center for American Progress. "The Power of the President: Recommendations to Advance Progressive Change" (Forward by John D. Podesta). November 15, 2010.

Cooper, Phillip J. 2002. *By Order of the President.* Lawrence, KS: University Press of Kansas.

Dombeck, Michael P., Christopher A. Woodand Jack E. Williams. 2003. *From Conquest to Conservation.* Washington, DC: Island Press.

Durant, Robert F. 1992. *The Administrative Presidency Revisited: Public Lands, the BLM, and the Reagan Revolution.* Albany: SUNY Press.

Epstein, David and Sharyn O'Halloran. 1999. *Delegating Powers.* Cambridge: Cambridge University Press.

Fiorino, Daniel J. 2006. *The New Environmental Regulation.* Cambridge, MA: MIT Press.

Fuchs, Edward P. 1988. *Presidents, Management and Regulation.* Englewood Cliffs, NJ: Prentice Hall.

Furlong, Scott R. 1997. "Interest Group Influence on Rule Making." *Administration and Society* 29(3):325–347.

Gottlieb, Robert. 1993. *Forcing the Spring.* Washington, DC: Island Press.

Graham, Mary. 1999. *The Morning After Earth Day.* Washington, DC: Brookings.

Guber, Deborah L. 2003. *The Grassroots of a Green Revolution.* Cambridge, MA: MIT Press.

Howell, William G. 2003. *Power Without Persuasion: The Politics of Direct Presidential Action.* Princeton, NJ: Princeton University Press.

Kagan, Elena. 2001. "Presidential Administration." *Harvard Law Review* 114: 2245–2385.

Kerwin, Cornelius M. 2003. *Rulemaking: How Government Agencies Write Law and Make Policy*, 3rd edition. Washington, DC: Congressional Quarterly Press.

Kingdon, John. 1995. *Agendas, Alternatives and Public Policies*, 2nd edition, Harper Collins.

Klyza, Christopher McGrory and David Sousa. 2008. *American Environmental Policy, 1990–2006.* Cambridge, MA: MIT Press.

Kraft, Michael E. 2000. "U.S. Environmental Policy and Politics: From the 1960s to the 1990s". *Journal of Policy History* 12(1):17–42.

Kraft, Michael E., Mark Stephan and Troy Abel. 2011. *Coming Clean.* Cambridge: MIT Press.

Layzer, Judith A. 2012. *Open for Business: Conservatives' Opposition to Environmental Regulation.* Cambridge, MA: MIT Press.

Leshy, John. 2001. "The Babbitt Legacy at the Department of the Interior: A Preliminary View." *Environmental Law* 31(2): 199–228.

Mayer, Kenneth. 2001. *With the Stroke of a Pen.* Princeton: Princeton University Press.

Moe, Terry M. 1985. "The Politicized Presidency" in *New Directions in American Politics*, edited by John E. Chubb and Paul E. Peterson. Washington, DC: Brookings.

Nathan, Richard P. 1975. *The plot that failed.* New York: Wiley.

Nathan, Richard P. 1983. *The Administrative Presidency*. New York: Wiley.

Neustadt, Richard E. 1991. *Presidential Power and the Modern Presidents: The Politics of Leadership from Roosevelt to Reagan*, 4th edition. New York: Free Press.

Repetto, Robert (ed.). 2006. *Punctuated Equilibrium and the Dynamics of U.S. Environmental Policy*. New Haven, CT: Yale University Press.

Rosenbaum, Walter A. 2012. *Environmental Politics and Policy*, 8th edition. Washington, DC: Congressional Quarterly Press.

Rosenbloom, David H. 2000. *Building a Legislative-Centered Public Administration*. Tuscaloosa, AL: University of Alabama Press.

Shabecoff, Phillip. 2001. *Earth Rising: American Environmentalism in the 21st Century*. Washington, DC: Island Press.

Shellenberger, Michael and Ted Nordhaus. 2004. "The Death of Environmentalism: Global Warming Politics in a Post-Environmental World." www.thebreakthrough.org/articles/the_death_of_environmentalism.

Shull, Steven. 2006. *Policy by Other Means*. College Station, TX: Texas A&M University Press.

Szasz, Andrew. 1984. "Industrial Resistance to Occupational Safety and Health Legislation: 1971–1981." *Social Problems* 32(2):103–116.

Tiefer, Charles. 1994. *The semi-sovereign presidency: the Bush administration's strategy for governing without Congress*. Boulder: Westview Press.

Trump, Donald. "State of the Union Address". January 30, 2018. www.presidency.ucsb.edu/documents/address-before-joint-session-the-congress-the-state-the-union-25.

Turner, Tom. 2009. *Roadless Rules: The Struggle for the Last Wild Forests*. Washington, DC: Island Press.

Warber, Adam. 2006. *Executive Orders and the Modern Presidency: Legislating from the Oval Office*. Boulder, CO: Lynne Rienner Publishers.

West, William F. 1982. "The Politics of Administrative Rulemaking." *Public Administration Review* 42(5): 420–426.

2

GREENING THE ADMINISTRATION

There was an unexpected confirmation fight when President Obama nominated Cass Sunstein as his first director of the Office of Information and Regulatory Affairs (OIRA). After a six-month delay and two holds on his nomination, Sunstein was confirmed by the narrowest margin of any previous OIRA nominee, and he went on to serve as "Regulatory Czar" for most of Obama's first term.[1] The office was a frequent subject of controversy ever since its creation, when it played a major role in Ronald Reagan's deregulatory agenda. Reagan empowered its director to conduct economic analyses of agency regulations, and subsequent presidents have continued to rely on centralized regulatory review to maintain White House control over policy development. Congress often clashed with presidents when it appeared that the office was improperly influencing agency decisions and undermining legislative intent. A law professor who had served in the Reagan Justice Department and written extensively on the value of economic analysis, Sunstein seemed to embody qualities that critics of regulatory review distrusted the most.[2] Most surprising about the confirmation battle was the substance of the objections. Senators raised questions about his personal views on a wide range of issues, from privacy to gun rights to animal rights. The proper role of regulatory review was not at issue, neither was White House influence over the rulemaking process. By then, it was understood that the White House had an interest in overseeing policymaking at the agency level and the authority to do so.[3]

Regulatory review has become a primary instrument for presidential management of environmental policy. The practice dates back to Nixon's "Quality of Life" reviews, later becoming a tool for subsequent administrations seeking to reduce the impact of regulations on key industries and limit the growth of the regulatory state. For this reason, Reagan aide Martin Anderson called it a "potent policy chokepoint."[4] This feature of the contemporary administrative presidency has also been used effectively to advance environmental goals during the Clinton and Obama years. Regulatory review has been called the eyes and ears of the president within the regulatory process. This oversight gives presidents significant ability to affect policy outcomes.

The President as Chief Executive

There is a broad literature about the various ways presidents influence policy. In a notable study that unified diverse perspectives on the presidency and the environment, Dennis Soden's overview of the "Environmental Presidency" examined this influence through the lenses of five roles of the president.[5] Two of these, commander-in-chief and chief diplomat, are roles which the president has traditionally exercised considerable power. At the other end of the continuum are the roles of legislative leader and party leader. The president occupies these roles but power is severely constrained because success depends upon the cooperation of numerous actors within the political system. At the center of this continuum is the role of chief executive. At times, each of the facets of the presidency has been instrumental to advancing environmental priorities. The present chapter argues that the role of chief executive, in particular, is the most critical as a means to pursue environmental policy change in an era of gridlock.

Presidents have a significant advantage over other institutions to set the policy agenda.[6] Because of the nationwide constituency, presidents have broad latitude to identify problems of national importance and to elevate issues on the agenda. This gives them significant power throughout the policymaking process. In the agenda-setting process, actors who succeed at defining a problem gain influence because they shape the terms of the debate and define the alternatives.[7] The likelihood of a policy response sometimes depends upon how a problem is learned about. Explanations of agenda setting usually focus on communication between decisionmakers within government and outsiders. Charles O. Jones defined models of the process where problems are brought to the attention of government officials by interest groups, and where officials within government define problems and encourage groups to mobilize their members and pressure other decisionmakers.[8] These are the familiar descriptions of the early environmental justice protests of the 1980s in North Carolina, and the Earth Day Demonstrations of 1970 (planned and coordinated by Senator Gaylord Nelson and a nationwide network of activists). Jones also describes a third model, in which experts within government identify a problem and work with private interests and decisionmakers simultaneously to raise awareness and build consensus for action. This better describes the early stages of contemporary environmental policymaking. Environmental bureaucracies are typically cast in terms of their obligation to implement policy. Agencies can use their knowledge to shape the agenda as well as respond to it.[9] Because emerging environmental problems are less salient and more technical in nature, scientists and regulators are more often the ones who discover a problem and recommend a course of action. Agencies can bring their concerns to Congress, which could lead to hearings, authorization for further study or eventually, a law. However, if the president is receptive to the agency's concern, there can be immediate action through any of a number of tools of administrative policymaking.

The Administrative Strategy

The environment has never been more than a second-tier issue for recent presidents.[10] Consequently, it is rarely worth the risk of subjecting their policy proposals to the legislative process. There are opportunity costs for pursuing sponsorship and votes for an environmental bill that could have been used for other priorities. Moreover, the president's reputation suffers if a White House-backed bill fails. Beyond the avoidance of legislative inertia or outright obstruction, there is an incentive to pursue environmental priorities through administrative channels when authority has already been delegated to the bureaucracy. Presidents have the ability to influence the direction of policy during implementation because they appoint top administrators of federal agencies, and are unlikely to select appointees who do not share the president's views about the direction of the policies they will carry out. Leadership of the bureaucracy has given the president a wider channel to affect policy outside the formulation and enactment of legislation.

Although the resources of the office allow the presidency to dominate every stage of the policy process, it is the president's role as chief of the executive bureaucracy that presents the most opportunities to shape policy. Unilateral powers, such as directives, executive orders, signing statements, and executive agreements, are used frequently to advance a range of policy goals. As this volume argues, Presidents Clinton, George W. Bush and Obama have pursued their environmental priorities with the help of the bureaucracy through the appointments, rulemaking, and regulatory oversight. All three recent presidents showed a preference for this approach to environmental policymaking, rather than the traditional strategy of working with Congress.

Not only does the executive bureaucracy have significant influence over policy implementation because of its rulemaking responsibility, but it is also a major actor throughout the policy process. In the formulation stage, for example, Congress and the president seek input from agencies because of their expertise. Agencies often have the final word when they translate statutory law into administrative rules because of the discretion Congress granted when it passed the Administrative Procedure Act (APA). The 1946 law effectively yielded the implementation decisions for major legislation to the bureaucracy by delegating the responsibility for the details of rules. In creating a procedure for writing regulations, the APA required agencies to create an opportunity for deliberation. Agencies are required to solicit feedback on proposed rules, and then consider and respond to the comments. However, the APA does not specify how they should use the information they receive.[11]

Agencies have wide latitude to design rules, because legislation consists of vaguely-worded goals forged by compromise. In order to appeal to a large enough coalition, specific provisions may be eliminated or avoided and left to the administrative agency. Therefore, this ambiguity is by design. Beyond attracting majority support, the challenge for authors of legislation is to craft legislation with the optimal balance between prescription and discretion.[12] Clear directions might help ensure that agencies will comply with Congressional intent, but legislators may lack the knowledge and expertise to predict how their policies might apply under specific circumstances. Deborah Stone argues that flexible rules allow agencies to respond creatively to new problems and particular situations.[13] Therefore, overly specific policies may prove to become as problematic as ambiguous directives.

Bureaucratic discretion was enhanced because of the ambiguous language of the environmental laws that Congress enacted in the 1970s. All regulatory policies are not alike, even though they share the purpose "to govern economic activity and its consequences at the level of the industry, firm, or individual unit of activity."[14] The environmental statutes are examples of social regulation, which is distinct from old-style economic regulation in several ways. First, economic regulation is justified by the Interstate Commerce Clause of the Constitution and is concerned with maintaining functioning markets for particular industries. In contrast, social regulation is justified by the Constitutional power to promote the general welfare and governs behavior across multiple economic sectors.[15] The movement to develop site-specific regulations is an attempt to reduce the conflict associated with industry-wide rules. Second, the general public is indifferent to regulations for specific industries, whereas social regulatory policies are enacted at the urging of public interest groups and enjoy widespread popular support. Although many of the environmental statutes contained detailed provisions and strict deadlines, goals were defined broadly and decisions about the means to achieve them were often left up to administrative agencies.[16]

Even though public opinion does not register the same intensity of support for environmental protection as in the era of Congressional activism, the nation's environmental protection policies remain popular. Thus, it is not politically feasible for policymakers to advocate rolling them back.[17] When an anti-incumbent tide washed over Congress in 1994, the Republican majorities that were swept in found it difficult to deliver on their anti-regulatory agenda. Talk of rewriting the Endangered Species Act and new "Takings" legislation faded, and Congress ultimately cooperated with the White House to reauthorize the Safe Drinking Water Act and pass the Food Quality Protection Act. As Chapter 3 explains, portions of the Congressional environmental policy agenda were enacted as a series of riders. Due to the salience of environmental issues, opposing interests continue to maintain cross-pressure within Congress.[18] Even when George W. Bush became president in 2001, temporarily ending divided government, lawmakers were reluctant to repeal statutes that they opposed, such as NEPA.

Whether the motive is to maintain control over a policy initiative by operating within the bureaucracy or simply to bypass Congress, administrative policymaking is most likely to succeed if the initiative has popular support and is consistent with the authority previously delegated by Congress. Despite fierce opposition in the Mountain West, Bill Clinton's conservation initiatives were popular nationally. George W. Bush had mixed success with rule changes to increase timber harvesting by reframing the issue as forest fire prevention, and relaxing pollution standards by stressing utility prices and economic growth. In these examples, both presidents attempted to appeal to the national constituency of the office, while painting their Congressional rivals as defenders of special interests.

Building Institutional Capacity

The modern institutional presidency is the product of reforms and innovations that occurred alongside the growth of the regulatory state. In order to maintain control over the direction and implementation of domestic policy, the Nixon White House developed a parallel policy structure to rival that of Congress. This policy establishment evolved considerably by the time Bill Clinton modified it for his own purposes. Richard Nixon was in many ways the architect of the modern institutional presidency. In a series of administrative reorganizations, Nixon created an establishment within the Executive Office of the President to develop policy proposals and promote his priorities across the executive branch.[19] The Office of Management and Budget (OMB), established to replace the Bureau of the Budget, would later become a central policy actor during the Reagan years. Nixon also created the cabinet-level Office of Policy Development and the Office of Assistant to the President for Domestic Policy. Nixon was able to exploit this permanent domestic policy presence in the White House to centralize agency activities for the benefit of his administrative presidency.[20] Recent presidents have continued to rely on this policy establishment to develop their programs and ensure that agency actions were consistent with presidential priorities. As they used these institutional arrangements to their advantage, presidents adding to the foundation left by their predecessors in the process. After a few more transitions of power, this structure evolved into a centralized organization that would come to dominate environmental policymaking.

President Carter paved the way for a major expansion of presidential control over administration by sponsoring government reform and institutionalizing regulatory review. Ironically, Carter's presidency began with a well-intentioned return to cabinet government.[21] Cabinet secretaries were empowered initially to name their own deputies and set the agencies for their departments. However, Carter's enthusiasm for cabinet government soured.[22] As the energy crisis and economic recession persisted, Carter empowered his Chief of Staff to coordinate policy initiatives across the administration. The reform bill that Carter signed into law created new patronage positions at the top and middle levels of federal agencies. The expectation was that increasing the number of political appointees would improve prospects for the president's agenda, since their loyalties would be with the president rather than the secretary. The Civil Service Reform Act of 1978 was passed with the stated purpose of centralizing White House control of executive agencies. Although Carter never had the opportunity to use them, the tools created by the law were later bequeathed to Ronald Regan, who used them to centralize policymaking within the White House. Surpassing the coordination powers of earlier presidents, Reagan was able to place loyalists deep inside the bureaucracy. Reagan began his presidency determined to halt the growth of the regulatory state and reverse domestic policy initiatives. This demanded an unprecedented degree of control over the organs of the state. To that end, the Office of Policy Development was given greater responsibility for coordinating the president's agenda across the federal government. Reagan also established "Cabinet Councils" that met regularly during his first term, taking the opposite approach that his predecessor had tested but abandoned in favor of a more centralized administration.

Regulatory Review

Policy coordination by the White House was further enhanced with the practice of centralized regulatory review. By layering new procedural requirements on top of the rulemaking process, presidents have accumulated greater influence over agency rulemaking over time. Although the Ford Administration ordered inflation impact analysis on an ad-hoc basis, this practice became systematic when Carter ordered federal agencies to perform regulatory analyses of proposed rules. This order established the Regulatory Analysis Review Group (RARG) to oversee agency reviews. This standing panel also conducted its own reviews of agency rules at the direction of the White House.

Centralized regulatory review became institutionalized in the Reagan Administration. Rather than allow agencies to continue conducting their own regulatory reviews, Reagan assigned this responsibility to OMB, the only organization within the executive branch with the resources to carry them out.[23] After initially dissolving RARG, Reagan embraced the idea of centralized regulatory review and institutionalized the process. His term in office began with an ambitious review of federal regulations, and with EO 12291, Reagan formalized the practice of cost-benefit analysis that was carried out only on a case-by-case basis under Carter. The Paperwork Reduction Act of 1980, signed into law by his predecessor, created OIRA within OMB. Reagan designated OIRA as the central clearinghouse for agency rulemaking and empowered the new office to guide policy development at the agency level. The centralized regulatory review process, located within the Executive Office of the President since 1981, is designed to ensure agency rules reflect the preferences of the president.[24]

By the end of the Reagan years, there was a great deal of mistrust between the administration and Congressional Democrats, who accused the White House of delaying and interfering with regulations under the EPA's Superfund program.[25] Congress cut OIRA's 1989 budget to curtail what they perceived as the administration's undue power to undermine policy implementation. Furthermore, when the senate refused to confirm President George H.W. Bush's nominee to lead OIRA, the White House responded by moving its regulatory oversight functions to the president's Council on Competitiveness, a little-known advisory committee chaired by Vice President Quayle. The council consisted of executive director Alan Hubbard and six cabinet-level officials, not including the EPA Administrator. Initially, the council was empowered to resolve disputes between agencies without requiring the president's involvement. It also held private meetings where regulated industries appealed for relief from regulatory excess.[26] The council's role in reviewing agency regulations began in 1991. When the EPA tried to reduce haze in the Grand Canyon by seeking a 90 percent reduction in sulfur dioxide emissions from the nearby Navajo Generating Station, the Council intervened. By going to the Council, the owners of the plant were able to persuade the EPA to propose an alternative 70 percent reduction as a compromise.[27] Through the council, the Bush Administration continued the centralization of the regulatory process and it also became useful as a point of interest group access.[28]

Since OIRA was defunded, Congressional Democrats were predictably scornful of the Quayle Council, which appeared to wield the same power with even less accountability. Their criticism centered on the panel's preoccupation with health and safety issues, its penchant for secrecy, and the privileges it appeared to offer industry groups.[29] The special access troubled lawmakers because it gave regulated industries another opportunity to win concessions they were unable to obtain through the open deliberations of the legislative process.[30] The Council on Competitiveness was abolished when Bill Clinton became president, yet he saw the value of centralized regulatory review and maintained the practice for his own ends. The process it created has endured, as both the essential structure of OMB and the practice of centralized clearance have remained intact through changes of presidents and parties.

Administrative Reform

The 1990s represented a new era for environmental politics because of a role reversal between the executive and legislative branches. Even though Clinton enjoyed large Democratic majorities in the 102nd Congress, they refused to act on his more ambitious objectives, from energy conservation, to natural resource policy reform, to a planned executive branch reorganization that would have elevated the EPA to a cabinet-level department. After 1994, Congress was less congenial toward environmental goals, and the presidency became more assertive.[31]

A series of executive orders signaled new policy priorities for his administration. For example, as discussed in Chapter 5, a 1993 order made the federal government subject to the reporting requirements of the Emergency Planning and Community Right-to-Know Act. Other executive orders created new institutions that would further centralize policymaking within the White House. Following through with his predecessor's commitments at the 1992 Rio Summit, Clinton established the President's Council on Sustainable Development (PCSD). In 1994, he created an interagency council on environmental justice by executive order. He enhanced the role of the CEQ and created the Office of the Federal Environmental Executive (OFFE) within the organization to coordinate policy initiatives across federal agencies. One of his second term executive orders established the Children's Public Health Advisory Committee to study the environmental, health, and safety effects of proposed regulations on children and recommend policy changes.

Clinton relied on appointments, administrative reorganization, and regulatory oversight to achieve his goals in the face of Congressional opposition, as did Reagan. Unlike Reagan, he used these tools of the administrative presidency in the service of environmental protection and reform.[32] Agency leaders with strong environmental credentials were appointed, including Bruce Babbitt as Interior Secretary, Carol Browner as EPA Administrator, and Mary Nichols as head of the agency's Office of Air and Radiation. However, his method of asserting control over the bureaucracy depended more upon the reforms he pursued and his management style. Whereas Reagan used the power of appointment to increase the number of patronage positions to help press the president's agenda within the bureaucracy, Clinton reduced the number of appointees. The National Performance Review, conducted in his first year in office, served to strengthen presidential administration because it helped earn the trust of career staff, who in turn were permitted to run their agencies with greater discretion.

Administrative reorganizations created new panels within the EOP that enhanced presidential control over policy. He split the Office of Policy Development in two, creating the National Economic Council and the Domestic Policy Council (DPC) which became the principal organizations for coordinating economic and policy across federal agencies.[33] At the same time, he streamlined the administration by removing advisory committees that were nonessential to the White House policy establishment. The Federal Advisory Committee Act was a tool that Congress used to maintain influence over federal agency agendas.[34] Clinton dissolved a number of these committees in his first term, which allowed him to take credit for reducing the size of government, but also helped consolidate his control over the executive branch.

The success of the administrative reform agenda was mixed, in part because the initiatives were unpredictable. For example, research on the Department of Defense found administrators were more likely to embrace reform when it supported their own priorities, and the enhanced discretion they gained was sometimes used to undermine administration goals, such as the remediation of toxic contamination. In this sense, there was a risk of administrative reforms becoming "weaponized" for inter-agency battles.[35] Many of the initiatives were short-lived because they were no longer supported by the White House following the Bush transition. Although Clinton's successor continued some of the nonregulatory approaches, other reinvention projects aimed at improving performance at federal agencies were discontinued.

Regulatory Activism

Clinton's regulatory activism was possible because of the way he reinvented the administrative presidency. Former administration official Elena Kagan attributes the success of his administrative strategy to the skillful use of regulatory review, presidential directives and appropriation of policy initiatives. Combined, these management techniques allowed Clinton to use the regulatory activity of the bureaucracy as an extension of his own policy agenda.

Rather than reverse the trend toward centralized regulatory review, Clinton preserved the process that he inherited from Reagan and Bush, adapting it for a pro-regulatory agenda.[36] He replaced Reagan's Executive Order with his own directive, E.O. 12866, which reaffirmed its key provisions, but with notable differences. Clinton expanded its scope to include independent agencies. The 1993 order continued the requirement that agencies submit major rules for review, and also specified that OMB conduct cost-benefit analyses of proposed regulations "to the extent permitted by statute." The order also included language to enhance accountability. It limited the time allowed for reviews, which precluded the process from being used as a delay tactic, and specified that only the director of OIRA could receive outside communication regarding the rules during the review.[37]

In another technique to manage the federal government's regulatory agenda, Clinton began the practice of issuing directives for agencies to engage in rulemaking for purposes within their discretionary authority. There were 107 directives issued to the heads of federal agencies during Clinton's two terms, compared to just nine in Ronald Reagan's two terms and four in George H.W. Bush's single term.[38] In doing this, he "added to the system of presidential oversight that he had inherited a scheme for direct presidential intervention."[39] Because rules carry the force of law and are resistant to change, agency rulemaking is an additional pathway for the president's agenda. Since the early 1990s, rulemaking has been considered a viable strategy for presidents when their ENR policy initiatives become bogged down in Congress.[40] Even when incoming presidents oppose rules promulgated by their predecessors, they are rarely willing to divert resources from their first hundred-days agenda to advocate administrative withdrawal or replacement by initiating a new rulemaking process. In this sense, there is a strong temptation for outgoing presidents to engage in "midnight" rulemaking to entrench their policy preferences.[41]

The forthcoming chapters describe conditions when legislation stalled, and Clinton directed agencies to make policy changes regarding land management, water quality standards, hazardous waste mitigation and greenhouse gas emissions.

In those cases, directives prompted agencies to take actions that had been author-ized by statutes. The flipside of this method is a formal refusal to implement pro-visions of a statute, a prerogative that Clinton asserted when he issued signing statements to reject provisions of bills without resorting to the veto. Although presidential signing statements date back to the 19th century, Reagan was the first to attach them to legislation as a way to thwart provisions that conflicted with his priorities. In 1993, Solicitor General Walter Dellinger prepared a memorandum for the White House counsel, arguing that signing statements could be used to declare that provisions of a bill would be unconstitutional as written.[42] Therefore, Dellin-ger advised, the president is justified in announcing an alternate interpretation at the time of the signing, in order to avoid such a conflict. This gave Clinton the legal justification that he sought. He later used a signing statement to fend off a legisla-tive assault on the Endangered Species Act, long before President George W. Bush was roundly criticized for enlisting the technique to serve his agenda.

The third aspect of Clinton's strategy was appropriation, demonstrated by his willingness to take ownership of agency actions. Clinton was far more open to policy proposals from the civil service, which could become administration prio-rities if they received his support. Whereas agency rulemaking would typically occur in relative obscurity, Clinton campaigned for his administrative policy initiatives, just as if he was "going public" to pressure lawmakers into supporting legislation. He often made personal announcements of proposed rules and final rules, wasting no opportunity to build public support and discourage legislative and legal challenges. As discussed in Chapter 3, once he decided to take owner-ship of a proposal from the Forest Service to ban new road construction in roadless areas, Clinton used a dramatic photo opportunity to announce the pro-posal, and another public appearance to announce the final rule 18 months later. Appropriating the strategy of "rhetorical presidency," Clinton used his public appearances and speeches to persuade the public and policymakers to support his environmental initiatives.[43] The difference, according to Kagan, was that he used his rhetoric for policies that originated within the administration:

> Nothing was too bureaucratic for the president....He emerged in public, and to the public, as the wielder of "executive authority" and, in that capacity, the source of regulatory action. As a result, during the Clinton years, the public Presidency became unleashed from the merely rhetorical Presidency and tethered to the administrative Presidency instead.[44]

Thus began Clinton's environmental presidency, which was to continue to employ administrative power to more rapidly affect change in environmental regulation.

In another initiative of the Clinton White House, agencies were encouraged to test alternatives to traditional regulation. Three major categories of alternative policy instruments were promoted. One approach is to customize policies. This includes negotiated rulemaking, which expanded after the National Performance

Review. The EPA began collaborating with stakeholders to develop pollution rules long before it was adopted widely by other agencies. This category also includes such instruments as performance standards and alternative standards, which increase flexibility and compliance options. A second category consists of market-based policies. These range from emissions trading markets to tax credits and subsidies to encourage conservation. These initiatives were fueled by a desire to foster cooperation with industry, which produced such collaborations such as the Partnership for a New Generation of Vehicles, which Clinton forged with the big three U.S. automakers in 1993. In that partnership, the federal government shared information with automakers in an effort to reduce research and development costs for more fuel-efficient vehicles.[45] Third, information technology has been deployed effectively to facilitate rulemaking and compliance with regulations. The EPA was an early adopter of electronic rulemaking, creating online dockets in the 1990s, and setting up a model for the federal e-rulemaking act of 2003. Moreover, the internet has enhanced the effectiveness of policies based on information provision, such as the Toxic Release Inventory. As the following chapters will show, presidential directives have been used effectively to advance a new generation of policies.

Some critics have questioned whether regulatory negotiation is any more effective than traditional regulation. One concern is that the process creates new pathways for conflict.[46] In addition, the very nature of negotiation encourages parties to downplay points of disagreement while focusing on areas of agreement in order to achieve consensus through compromise. For example, Cary Coglianese found that CSI projects were more likely to produce ineffective and noncontroversial proposals, such as voluntary standards and support for research, while avoiding tough decisions. Rather than defusing conflicts, these authors argue that conflicts are more likely to be displaced, only to re-emerge during implementation. Other expected benefits of leadership programs have proven difficult to observe and measure; this is because, "while environmental goals are themselves not always easy to define," as Borck et al., ask, "for the most part they are the consequences of economic outputs and can be operationalized using outputs that can be isolated, measured and tracked....But what, exactly, does it mean to 'improve relationships' among facilities, agencies and communities? Or to 'change culture' or 'enhance trust?'"[47] Finally, alternative policy tools raise legitimate concerns about democratic accountability. Theodore Lowi, a critic of delegating legislative authority, likened the rise of the regulatory state as "policy without law" because significant implementation decisions are made without the opportunities for democratic deliberation and representation that are characteristic of the legislative arena.[48] Creating a pathway for regulated interests to bypass legislative rules and negotiate their own arrangements means that policy is twice removed from the legislative process.

Clinton also advanced environmental policy goals through diplomacy. Unwilling to risk senate rejection of environmental treaties, he relied instead on executive agreements to express U.S. commitments. As explored in Chapter 6,

Clinton signed the Kyoto agreement in 1997 but decided not to submit the treaty for senate ratification. In 1998, the U.S. signed the Rotterdam Convention on Hazardous Chemicals as well as the Protocol on Persistent Organic Pollutants. Neither treaty was sent to the senate for ratification, which would also have required Congress to approve amendments to TSCA and the Federal Insecticide, Fungicide and Rodenticide Act. Like these agreements, many of Clinton's administrative initiatives required partisan continuity for policy continuity. Without Congressional action or a Gore White House to follow through, they stalled during the Bush/Cheney years.

Policy Retrenchment in the Bush Years

Just as Clinton had transformed the administrative presidency into a force for regulatory activism, the George W. Bush White House made easy work of redirecting it toward the functions it had in the Reagan years. Building onto structures created during Clinton's presidency, Bush further centralized the regulatory agendas of federal agencies to ensure that the entire government shared his agenda of deregulation and development of energy and natural resources. This was nothing unique to environment and natural resource policy; rather it was an expression of the White House strategy of slowing and containing the regulatory state. Appointments were instrumental to these objectives. According to OIRA Director John Graham, "Bush looked for candidates who supported the need to approach federal regulation from a cost-benefit perspective."[49]

Bush made sophisticated use of appointments as a centralization strategy, with the largest increase of political appointees within agencies that were seen as opposing the priorities of the White House.[50] Vice President Cheney was given latitude to choose the administrators of energy, environment and natural resource agencies, and industry representatives were named to prominent positions at the top and secondary levels. Jeffrey Holmstead, a lobbyist who represented utility companies, was named Assistant Administrator for Air and Radiation at the EPA. As Deputy Administrator, they named Linda Fisher, whose resume included the EPA and Capitol Hill, as well as the chemical company Monsanto. At the Justice Department, Bush and Cheney's choice for Assistant Attorney General in charge of the Natural Resource Division was Thomas Sansonetti, a coal lobbyist from Wyoming. Gayle Norton, a protégé of Ronald Reagan's controversial Interior Secretary James Watt, was appointed to the same position in the Bush/Cheney Administration. To lead the U.S. Forest Service, they appointed Mark Rey, a timber lobbyist who advocated opening national forests for oil and gas drilling and additional logging as well.[51]

One political appointee whose background did not resemble the industry insiders appointed to sub-cabinet level offices was Bush's first EPA Administrator, former New Jersey Governor Christine Todd Whitman. Although she was confirmed in a unanimous Senate vote and began her term by pledging continuity with many Clinton-era policies, the new administration changed course on several fronts and Whitman was marginalized. When Vice President Cheney convened a task force to develop national energy strategy, Whitman disagreed with the decision to exclude environmental groups from the meetings. She also took issue with the group's objective of using administrative channels to modify the New Source Review (NSR) requirements of the Clean Air Act, arguing instead that any such reform should be left to Congress as part of the "Clear Skies" legislation that she supported.[52] After more than two frustrating years at the EPA, Whitman resigned.

Bush relied on rulemaking for most of his environmental policy changes he sought, issuing directives to the heads of agencies to develop rules under their existing authority. During his first term, OIRA assumed a more active role in the regulatory process. Expanding beyond regulatory review, the White House took the unprecedented step of allowing OIRA to offer its own suggestions for regulatory changes.[53] Eight of the 15 so-called "prompt letters" letters concerned environmental policies, whereas the others directed agencies to conduct rulemaking related to health, safety, and banking regulation.[54] Table 2.1 lists the letters that prompted environmental rulemaking. Some urged rulemaking to increase the efficiency of existing programs, such as expanding electronic reporting for the Toxic Release Inventory (TRI). Others were intended to identify opportunities to improve upon existing regulatory practices. For example, the Agriculture Department was encouraged to consider voluntary and market-based approaches to promote environmental stewardship as alternatives to command-and-control policies.

Bush initially left Clinton's procedures for regulatory review intact. Following the Democrats' takeover of Congress during his second term, Bush issued E.O. 13422 to consolidate presidential control of the executive branch. Now that the White House was forced to share oversight with an opposition Congress, the already-centralized structure inherited from the Clinton Administration was no longer enough. The 2007 order revoked Clinton's 1993 order, replacing it with a new regulatory review procedure that required agencies to explicitly define the policy problem that justified the regulation, and identify a specific market failure that the proposed regulation would be intended to solve. It also required agencies to designate a political appointee within the agency as a regulatory compliance officer, whose approval is required before any proposal can be listed on the semiannual regulatory agenda.

The Bush Administration used a practical strategy to reverse course on many of the Clinton Administration policies it inherited. Many of the regulations still under development were withdrawn or returned to the agencies. According to John Graham, OIRA returned more regulatory proposals during Bush's first year in office than during Clinton's two terms.[55] Rather than commit resources to a lengthy rulemaking process to reverse rules already in effect, the administration took advantage of litigation in the pipeline and used settlements as the basis for

TABLE 2.1 "Prompt Letters" Sent to Environmental Agencies, 2001–2006

Year	Agency	Issue
2001	EPA	Research for Targeted Particulate Matter Regulations
2002	EPA	Electronic TRI Reporting
2002	EPA	Nonroad Diesel Engines and low-Sulfur Fuel
2002	USDA	Environmental Quality Incentives Program
2003	Department of Energy	Energy Modeling with Electric and Hybrid Vehicles
2003	Department of Interior	Availability of Critical Habitat Mapping Data
2004	EPA	Standards for pathogen contamination of beaches
2006	EPA	Alternative Risk Modeling for Particulate Matter

the new policies. For example, the National Association of Homebuilders filed a lawsuit which challenged a critical salmon habitat designation by the National Marine Fisheries Service in 2000. Rather than fight the lawsuit, the administration settled, reopening the public comment period and ultimately reduced the size of the critical habitat by 80 percent. In effect, Bush was "getting his way by settling."[56]

By abandoning the reinventing government project, the Bush Administration largely dismantled the administrative reforms that comprised one of Clinton's major environmental policy achievements. However, one aspect of this legacy continued and contributed to Bush's environmental policies. The reinvention initiative of the 1990s spawned several public-private partnerships that allowed the EPA to experiment with voluntary and collaborative policies toward various industries. In an expansion of one Clinton initiative, Bush encouraged the formation of partnerships that emphasized voluntary environmental standards. The EPA formed 29 new partnerships during the Bush years, compared to 23 during the Clinton years.[57] Unlike the Clinton-era partnerships, which utilized a variety of alternative mechanisms for standard setting, monitoring and enforcement, the Bush-era partnerships emphasized voluntary programs at the exclusion of all other approaches. In this sense, the Bush Administration discarded much of the innovation that occurred previously, and partnerships became a substitute for regulation.[58]

In another practice characteristic of the Bush administrative strategy, the White House sought to prevent the states from enacting environmental standards stricter than those promulgated at the national level. A particularly contentious provision of the Clear Skies and CAIR proposals was the requirement that state-level restrictions on emissions sources be rescinded.[59] It is not uncommon for federal environmental laws to preempt states for the purpose of establishing floors. What was unusual about the Bush Administration's regulatory federalism was the insistence on setting ceilings that the state could not exceed.[60] In an unprecedented move, the EPA denied California a waiver request under the Clean Air Act to reduce carbon emissions.[61] In a twist on the politics of states' rights, the administration began inserting preemptive language in the preambles of regulations, sometimes after rules had been through the notice-and-comment process. Much of this language targeted federal Food and Drug regulations, as well as other health and safety rules, in what appeared to be a campaign to accomplish tort reform administratively.[62] This strategy of regulatory preemption raised concerns about transparency and agency capture, since preemptive rules tend to be industry-friendly.[63] The strategy also stands in marked contrast with other Republican presidents, who would frequently appeal to constituencies by proclaiming support for states' rights.

From Preemptive to Progressive Federalism

During the months that Cass Sunstein awaited his confirmation hearing, he actively shaped the administration's regulatory agenda and collaborated with the White House on new executive orders. Barack Obama issued several executive orders early in his first term that brought further changes to the administrative presidency. In one of the first, Obama revoked Bush's 2007 executive order that imposed new restrictions on regulatory agencies. Consequently, the regulatory review requirements reverted to the procedures established under Clinton's EO 12866, which was also in effect for the first six years of the Bush Administration. That the president would use this system as the basis for oversight of federal agencies indicated that he shared much of the belief in a unitary executive as his two predecessors.[64]

Although Obama never relinquished the centralized control of the federal government that his predecessor maintained, his approach to intergovernmental relations was extremely flexible when state-level initiatives supported his agenda. Several state governments were frustrated with the Bush Administration's resistance to their innovations in the environment and other policy areas. Obama's approach to "progressive federalism" was to allow states to experiment with their own policies as long as they matched or surpassed the performance of federal standards.[65] Shortly after taking office, Obama called for a review of all regulations with preemption statements. Obama's memorandum directed the heads of federal departments and agencies to review any rules "issued within the past ten years" to determine whether they were legally justified.[66]

As with his two most recent predecessors, Obama made careful use of appointments to build an administration whose environmental priorities were cohesive and consistent with his own. Thus, he announced he would return science to its rightful place in environmental policymaking. He won praise for creating a position of scientific integrity officer at the Department of Interior and appointed several officials with strong scientific credentials to serve during his first term. This group including Nobel Prize-winning physicist Steven Chu as Secretary of Energy, and Jane Lubchenco, a leading marine ecologist to lead

NOAA. In one of his first executive orders, Obama reconstituted the Domestic Policy Council to include as members the Director of the Office of Science and Technology Policy, John Holdren and Carol Browner, the former EPA Administrator who Obama named to the new position of Assistant to the President for Energy and Climate Change.[67]

Informally, Browner's position also came to be known as "Energy and Climate Czar." It was one of dozens of positions created to better oversee and coordinate action across the administration. Another was Todd Stern, informally known as "International Climate Czar," appointed to oversee climate talks in advance of the Copenhagen summit. Obama was hardly the first president to appoint czars; beginning with Nixon, who named a "drug czar" to coordinate the War on Drugs, every president had seen the value of naming a point-person who could cut through the bureaucracy and maintain a direct line of contact to the Oval Office for urgent policy priorities. Obama expanded upon this approach by naming a czar for nearly every domestic and international policy area on his agenda.[68] These included a Great Lakes Czar (Cameron Davis) and a Green Jobs Czar (Van Jones).

Obama's czars were unexpectedly controversial. At one time, there were as many as 29 separate policy czars, mostly employed as presidential advisors not subject to Senate confirmation. Critics on the right and the left raised questions about their powers, their positions in the hierarchy of the administration, and the absence of oversight that normally comes with the confirmation process. The controversy was an unwelcome distraction from Obama's first-year agenda, as the White House stumbled to form a coherent response to those questions. In the backlash to Obama's czars, Congress stepped up oversight, and debated cutting off funding for the new positions. Those presidential advisors were active in policy planning and coordination early in Obama's first term, but many left the administration after Congress thwarted Obama's legislative proposals following the 2010 election. Even the position of Assistant to the President for Energy and Climate was dissolved after Browner stepped down in 2011.

Deconstructing the Administrative State

Policy retrenchment in the Trump presidency went far beyond the Reagan and Bush-era obsession with repealing individual regulations. Trump and his advisors hinted about this, but the agenda was never articulated publicly until after his inauguration. At a February 2017 gathering of conservative activists, White House Chief Strategist Steve Bannon and Chief of Staff Reince Preibus announced the new administration would be dedicated to the "deconstruction of the administrative state." Bannon explained their commitment was not just to deregulation, but to undo the entire framework of regulations, taxes and trade deals they saw as fundamentally illegitimate and a threat to national sovereignty.[69]

Trump was extremely slow in staffing his administration, and dozens of key positions remained vacant for over a year. He had little interest in staffing regulatory agencies and OMB with seasoned veterans who could move the levers of the policy state to serve a coherent agenda, as Bush and Cheney had done in 2001. Rather, top positions were given to political supporters with little regard for their agencies' missions. Following the 2016 election, representatives of Trump's transition team did not bother to show up for the required briefings by the outgoing managers of federal agencies.[70] The chaotic transition was initially led by New Jersey Governor Chris Christie, who complied lists of candidates to staff the administration, but he was unceremoniously fired before the work was complete. Bannon, whose hostility to Washington insiders was well known, made a show of throwing Christie's binders of resumés into the trash.[71]

If Trump was not prepared to staff his own administration, others were willing to do it for him. Hundreds of names were submitted by the Heritage Foundation, which boasted their nominees were employed at every federal agency.[72] Other positions were given away as rewards to the motley coalition of business interests that supported his campaign. When he became a candidate for president, Trump was shunned by Chamber-of-Commerce-type corporate interests that typically fund Republican front-runners (those groups were backing such candidates as Marco Rubio and Jeb Bush). Those willing to meet with him in Trump Tower were mostly leaders of privately held corporations, including coal barons and oil-and-gas wildcatters. These individuals formed the Trump Leadership Council. Through their donations and support, they forged a policy agenda for a candidate who was otherwise not very ideological and had little interest in policymaking, other than the issues of immigration and trade at the center of his campaign announcement.

The council was organized by Harold Hamm, the owner of Continental, an oil and gas petroleum pipeline company. Hamm was instrumental to Trump's fundraising efforts at a time when most of corporate America was keeping its distance. Among the early supporters he brought to the council were leaders of coal companies such as Kraft Industries, Devin Energy, and Murray Coal, whose president Robert Murray would dictate much of Trump's policy agenda after the election. In his home state of Oklahoma, Hamm was the largest financial donor to Scott Pruitt, the state attorney general who filed a dozen lawsuits against the US EPA on behalf of the fossil fuel industry before he was tapped to lead the agency and curb its regulatory activity in the Trump Administration. Pruitt's deputy and successor as EPA administrator would be Andrew Wheeler, who was previously a lobbyist for Murray Energy. Pruitt and Wheeler were just two examples of Trump using his appointment power to steer the administration in an antiregulatory direction. At the Department of Interior and NOAA, top posts and sub-cabinet level positions were staffed with appointees who had previously worked for the chemical industry. Trump issued a directive for agency heads to form "deregulation teams" tasked with identifying rules to repeal.

Nominating judges to the federal bench is a significant presidential power, but it is imprecise and unpredictable for serving a policy agenda because presidents can rarely predict what cases will come before the courts and how judges are likely to decide. Trump had two early opportunities to name judges to the Supreme Court that seemed to act in accordance with his views on deregulation. As the Supreme Court drifted to the right, it has been increasingly willing to hear cases regarding delegation and rethink its traditional deference to agencies.[73] Trump nominees have continued this trend. Associate Justice Neil Gorsuch, Trump's first appointee, has indicated his openness to reconsider nondelegation, an old and defunct legal doctrine which was used to challenge the New Deal programs in the 1930s, before the Administrative Procedures Act standardized the process for agency rulemaking and established boundaries for judicial oversight. When the delegation power was recently upheld in the 2019 case *Gundy v. United States*, Gorsuch was not among the majority. In a dissenting opinion, Gorsuch wrote of the need to curb the power of Congress to delegate to federal agencies, adding "If a majority of this court were willing to reconsider the approach we have taken for the last 84 years, I would support that effort." Prior to his nomination to the Supreme Court, Judge Brett Kavanaugh had publicly expressed his opposition to "Chevron-style deference," a reference to the Supreme Court's decision in *Chevron v. NRDC* (1984) which is viewed by many environmentalists as a crucial firewall that keeps the courts from interfering with the EPA and other agencies on the substance of their regulations.

The Return of the Legislative Veto

Trump's policy agenda had some assistance from Congress for some of the initial heavy lifting. The outcome of the 2016 election was one-party rule, and the 117th Congress, which shared the administration's deregulation agenda, was in a unique position to rescind several Obama-era regulations that took effect in the previous six months. The 1996 Congressional Review Act (CRA) gave Congress the power to review and overturn rules before they went into effect by passing a disapproval resolution, with a simple majority vote in both houses, and no floor debate.

The CRA was controversial because it resembled a legislative veto, which was legally dubious. In 1983, the Supreme Court ruled in *Chadha v. INS* that the Constitution did not provide Congress with extra-legislative veto powers over executive actions, and the court struck down an immigration law which allowed Congress to overrule deportation decisions by majority vote in a single house. The *Chadha* decision did not completely abolish the legislative veto, but it changed the way Congress crafted legislation in order to preserve its power to overrule executive actions in specific circumstances.[74]

The authority grated in the CRA was more limited in scope than the legislative veto provisions that were ruled unconstitutional by the U.S. Supreme Court's *Chadha* decision. The CRA was intended to deter last-minute rulemaking by a lame-duck administration, in the event Clinton lost his reelection bid in 1996. Congressional Republicans eventually used CRA after four years to repeal OSHA's ergonomic standards, which were designed to prevent on-the-job repetitive motion injuries.[75] The final ergonomic standards were published just after the November 2000 election, and they took effect just four days before Bush's inauguration. OSHA began developing the ergonomics standards ten years earlier, during the administration of Bush's father, had spent more than ten years developing the ergonomics standards, but following the midterm election of 1994, the rulemaking was delayed by a series of legislative riders that prohibited the agency from completing its work until late in Clinton's second term. Prior to 2016, this was the only example of Congress using this power in the 20-year history of the CRA.

Of the 14 CRA actions taken in early 2017, four struck down regulations regarding energy and the environment. As discussed in the following chapter, one of the targeted regulations was the EPA's stream protection rule, which Robert Murray opposed. There were also rules which defined the planning process by the Bureau of Land Management, and regulations on hunting predators in Alaska. Finally, Congress used the CRA to rescind a rule requiring petroleum companies to disclose payments to foreign governments. Other CRA resolutions targeted transportation, education, occupational health and safety regulations, and even a rule that restricted the sale of firearms to mentally ill individuals. The unprecedented use of the CRA to rescind so many regulations allowed Trump to repeal much of Obama's regulatory legacy with the stroke of a pen.

Interstate Air Pollution: A Case Study in Frustration

The signature environmental initiative of George W. Bush illustrates the peril of using the regulatory process for an ambitious policy change. Initially, this was floated as the Clear Skies legislative proposal in 2002. The centerpiece of the bill was an emissions trading market to reduce pollution sulfur dioxide, nitrogen oxide and mercury pollution from stationary sources. Besides improving air quality, the legislation had the potential to resolve long-simmering conflicts over implementation of the Clean Air Act. The first was the problem of emissions that cross state boundaries and affect air quality in states downwind. The second was enforcement of the Clean Air Act's New Source Review (NSR) requirement.[76] In the Senate, the bill stalled in committee as it became entangled in debates over toxic pollution from mercury, and the insistence of several senators that any legislation include controls for carbon dioxide to combat global warming (see Chapter 6).

The components of Clear Skies were repackaged as a series of regulatory proposals. In Eisner's analysis, Bush turned to the administrative strategy as soon as his legislative initiative failed.[77] New rules were proposed to create cap-and-trade markets administratively after the defeat in Congress. The first was the Clean Air Interstate Rule (CAIR) which promised to reduce SO2 by as much as 70 percent and NOx by 60 percent from 2003 levels. As John Graham later wrote:

Rather than fold, Bush launched a series of clean-air rulemakings to achieve similar objectives. By relying on his executive authority to implement his clean-air agenda, Bush's policies became vulnerable to a complex series of lawsuits by national environmental groups, the regulated industries, and the states. The Bush Administration lost some important cases but won others.[78]

The rules creating emissions trading markets were examples of battles the administration lost. The CAIR was remanded in 2008, and its companion Clean Air Mercury Rule was vacated in 2005 (see Chapter 5). With the shift to an administrative strategy, Bush also changed policy on New Source Review. Pending enforcement suits were dropped when Clear Skies was announced. The EPA relaxed enforcement in 2003, later promulgating rules to allow power generators to extend the life of existing facilities by replacing equipment without meeting the higher emission standards required under NSR standards.[79]

In its December 2008 decision on the CAIR, the U.S. District Court in D.C. ruled that it violated the Clean Air Act because the rule authorized trading emission credits between programs. This forced the incoming Obama Administration to devote resources to promulgating a replacement rule, on top of its own priorities, to comply with a court order. Its replacement, the 2011 Cross-state Pollution Rule, was designed to accomplish most of the same goals as the one it replaced, but it addressed the court's concerns by placing more restrictions on participants in the cap-and-trade market. In 2012, the U.S. Court of Appeals in D.C. determined the trading market violated the Clean Air Act and vacated the replacement rule. In an opinion by Judge Brett Kavanaugh, the majority ruled that the EPA exceeded its authority by requiring states "to reduce their emissions by more than their own significant contribution to a downwind state's nonattainment." In a dissenting opinion, Judge Judith Rogers wrote the court "should defer to agency interpretations in areas where the Clean Air Act was silent." The court directed the EPA to rewrite the rule, unless Congress was to amend the Clean Air Act first.[80] The Federal government appealed the U.S. Supreme Court, and the Cross-state Pollution Rule was reinstated in 2014, finally closing the book on the Bush-era rule.

Conclusion

Every recent president has expanded the capacity of the office to influence environmental policy. Bill Clinton turned to the vast institutional presidency he received from Ronald Reagan and used that apparatus to advance his own agenda while fending off Congressional assaults on the regulatory state. He pursued policy changes through rulemaking, further centralized the regulatory review process, and reoriented major policies through his administrative reform initiatives and public-private partnerships. Finally, Clinton made aggressive use of unilateral action in the form of proclamations, executive orders, executive agreements and signing statements to bypass Congress.

Although the environment was not a major priority for his administration, George W. Bush pursued an expansive energy and resource development agenda. To these ends, he used the same administrative strategy as Clinton, albeit with greater assertiveness. The institutions and strategies that Clinton had built upon the foundation of the presidency that he inherited would prove instrumental to removing regulatory obstacles to Bush's policy agenda. Bush largely avoided legislation to roll back existing environmental policies, choosing instead to relax regulations through the rulemaking process and by expanding the type of voluntary agreements and partnerships begun under his predecessor. The administration settled lawsuits against natural resource agencies, using those settlements as the impetus to develop new, more industry-friendly rules. Furthermore, his administration bolstered the policy-by-rulemaking approach by giving OMB a more active role and by including preemption clauses in the preambles of new regulations. This discouraged states from responding with stricter standards of their own and minimized the risk of litigation over federalism disputes.

Faced with an uncooperative Congress and other political constraints, Barack Obama made full use of the institutions and arrangements of earlier presidents to pursue a broad range of environmental policy goals. Like Reagan, Obama staffed his administration with loyal appointees committed to his policy goals. From their positions in federal agencies, administration officials wrote rules to advance the president's priorities through the regulatory process, which was now centralized within the White House. From Clinton's playbook, Obama used unilateral actions including directives and executive orders to avoid the legislative process and negotiated executive agreements on the environment to avoid going to the senate to ratify treaties. He also turned to partnerships and voluntary agreements to pursue collaborative policies on pollution and conservation. Just as Bush made every advantage of the tools bequeathed to him by his predecessor, Obama's hand was strengthened by the precedents of the Bush years. Appropriating the strategies of his predecessor, the Obama Administration settled litigation and used those deals as starting points to initiate new procedures to reverse Bush-era rules and promulgated new regulations to advance policy goals that had languished during the voluntary experiments of the two previous presidents.

Republican Donald Trump pursued an administrative strategy from the outset, despite emerging from the 2016 election with solid GOP majorities in both houses of Congress. This placed the party in a strong position to pursue its agenda in the form of legislation, but the administration seemed to have little use for Congress beyond tax cut legislation. Virtually all of his other policy goals, from rescinding Obama's national monument designations to the coal industry's wish list, could be accomplished with greater efficiency through the rulemaking process. Pursuing these policy changes required little attention or effort from Trump himself. As the following chapter shows, it was enough to simply appoint industry lobbyists to key positions in the administration.

Notes

1 For more on the controversy over the Obama Administration's "policy czars," see Justin S. Vaughn and Jose D. Villalobos, 2015, *Czars in the White House*, Ann Arbor: University of Michigan Press.

2 Cass R. Sunstein, 2002, "A New Executive Order for Improving Federal Regulation? Deeper and Wider Cost-Benefit Analysis."

3 Stuart Shapiro argues that the bipartisan acceptance of Reagan's centralized regulatory review process was the result of its use in a Democratic Administration. By reaffirming regulatory review and continuing the practice, Bill Clinton legitimized its role. See Stuart Shapiro, 2007, "Presidents and Process: A Comparison of the Regulatory Process Under the Clinton and Bush (43) Administrations," *Journal of Law and Politics* 23(2): 393–418.

4 Martin Anderson, 1990, *Revolution: The Reagan Legacy*, 2nd edition, Stanford, CA: Hoover Institution.

5 Dennis L. Soden and Brent S. Steel, 1999, "Evaluating the Environmental Presidency" in *The Environmental Presidency*, edited by Dennis L. Soden, Albany: State University of New York Press.

6 Barry Glassner notes that presidents and the press frequently collude to raise the profile of certain problems such as illegal drugs and toxic waste. This sets the policy agenda, but also distorts public perceptions of relative risk. See Barry Glassner, 1999, *The Culture of Fear*, New York: Basic Books.

7 John Kingdon, 1995, *Agendas, Alternatives and Public Policies*, 2nd edition, Harper Collins.

8 Charles O. Jones, 1977, *An Introduction to the Study of Public Policy*, North Scituate, MA: Duxbury Press.

9 Daniel J. Fiorino, 1995, *Making Public Policy*. Berkeley: University of California Press.

10 Dennis L. Soden and Brent S. Steel, 1999, "Evaluating the Environmental Presidency" in *The Environmental Presidency*, edited by Dennis L. Soden, Albany: State University of New York Press.

11 Cornelius M. Kerwin, 2003, *Rulemaking: How Government Agencies Write Law and Make Policy*, 3rd edition, Washington, DC: Congressional Quarterly Press.

12 Deborah Stone, 2002, *Policy Paradox: The Art of Political Decisionmaking*, New York: Norton.

13 Ibid.

14 Marc A. Eisner, Jeff Worsham and Evan J. Ringquist, 2006, *Contemporary Regulatory Policy*, 2nd edition, Boulder, CO: Lynne Reinner.

15 Barry D. Friedman, 1995, *Regulation in the Reagan-Bush Era*, Pittsburgh: University of Pittsburgh Press.

16 Kenneth J. Meier, 1985, *Regulation: Politics, Bureaucracy and Economics*, New York: St. Martin's.

17 George C. Eads and Michael Fix, 1984, *Relief or Reform: Reagan's Regulatory Dilemma*, Washington, DC: Urban Institute.

18 Marc A. Eisner, Jeff Worsham and Evan J. Ringquist, 2006, *Contemporary Regulatory Policy*, 2nd edition, Boulder, CO: Lynne Reinner.

19 Paul C. Light, 1999, *The President's Agenda: Domestic Policy Choice from Kennedy to Reagan*, 3rd edition, Baltimore: Johns Hopkins University Press.

20 Richard P. Nathan, 1983, *The Administrative Presidency*, New York: Wiley.

21 The experience of the Nixon White House was still a vivid memory when Carter took office in 1977. Eager to build trust, Carter reassured the public that the White House staff would not be allowed to usurp the authority of cabinet officers. See Shirley Anne Warshaw, 1994, "The Carter Experience with Cabinet Government" in *The Presidency and Domestic Policies of Jimmy Carter*, edited by Herbert D. Rosenbaum and Alexej Ugrinsky, Westport, CT: Greenwood.

22 Hugh Heclo has noted a tendency for newly-elected presidents to affirm their belief in cabinet government, which he attributes to the desire to reassure the public that the new president is not power-hungry, rather than a concrete plan to manage his administration: "It is significant that no president has ever left office extolling the virtues of cabinet government." See Hugh Heclo, 1983, "One Executive Branch or Many?" in *Both Ends of the Avenue*, edited by Anthony King, Washington, DC: American Enterprise Institute.

23 Barry D. Friedman, 1995, *Regulation in the Reagan-Bush Era*, Pittsburgh: University of Pittsburgh Press.

24 Ibid.

25 Conrad B. MacKerron, "Has OMB Been Out of Bounds on EPA?", *Chemical Week* (March 22, 1989): 8.

26 Thomas O. McGarity and Sidney A. Shapiro, 1993, *Workers at Risk: The Failed Promise of the Occupational Safety and Health Administration*, Westport, CT: Praeger.

27 Victoria Sutton, 2003, "The George W. Bush Administration and the Environment," *Western New England Law Review* 25(1): 221–42.

28 Jeffrey Berry and Kent E. Portney, 1995, "Centralizing Regulatory Control and Interest Group Access: The Quayle Council on Competitiveness" in *Interest Group Politics*, 4th edition, edited by Allan J. Cigler and Burdett A. Loomis, Washington, DC: Congressional Quarterly Press.

29 Robert J. Duffy, 1996, "Divided Government and Institutional Combat: The Case of the Quayle Council on Competitiveness," *Polity* 28(3): 379–399.

30 In a 1992 hearing, House Oversight Committee Chair Henry Waxman accused the secretive council of inserting industry-backed provisions in to legislation after they had been rejected in the Congressional debate. See Waxman, Henry 2008. Opening Statement of House Committee on Oversight and Government Reform, Business Meeting Regarding Contempt Resolution (June 20) One Hundred Tenth Congress.

31 Christopher Bosso, 2005, *Environment Inc.*, Lawrence, KS: University of Kansas Press.

32 Norman J. Vig, 2006, "Presidential Leadership and the Environment" in *Environmental Policy: New Directions for the Twenty-First Century*, 6th edition, edited by Norman J. Vig and Michael E. Kraft, Washington, DC: Congressional Quarterly Press: 100–123.

33 The two councils inherited what had been Office of Policy Development's economic policy and domestic policy responsibilities. See James P. Pfiffner, 2005, *The Modern Presidency*, 4th edition, Belmont, CA: Thompson Wadsworth.

34 Steven J. Balla and John R. Wright, 2001, "Interest Groups, Advisory Committees and Congressional Control of the Bureaucracy," *American Journal of Political Science* 45(4): 799–812.

35 Robert F. Durant, 2007, *The Greening of the U.S. Military*, Washington, DC: Georgetown University Press.

36 Kelley argues that Clinton's role in creating structures that centralized executive power in the contemporary presidency is often overlooked by critics because it was used to such different ends as Reagan and George W. Bush. See Christopher S. Kelley, 2010, "The Unitary Executive and the Clinton Administration" in *The Unitary Executive and the Modern Presidency*, edited by Ryan J. Barilleaux and Christopher S. Kelley, College Station, TX: Texas A&M Press.

37 Elena Kagan, 2001, "Presidential Administration," *Harvard Law Review* 114: 2245–2385.

38 Ibid.

39 Ibid.

40 Klyza and Sousa argue that rulemaking has emerged as a particularly effective form of executive policymaking in gridlocked policy areas such as environment and natural resources. See Christopher McGrory Klyza and David Sousa, 2008, *American Environmental Policy, 1990–2006*, Cambridge, MA: MIT Press.

41 David M. Shafie, 2013, *Eleventh Hour: The Politics of Policy Initiatives in Presidential Transitions*, College Station, TX: Texas A&M University Press.

42 Christopher S. Kelley, 2010, "The Unitary Executive and the Clinton Administration" in *The Unitary Executive and the Modern Presidency*, edited by Ryan J. Barilleaux and Christopher S. Kelley, College Station, TX: Texas A&M Press.

43 Jeffrey Tulis, 1987, *The Rhetorical Presidency*, Austin: University of Texas Press.

44 Elena Kagan, 2001, "Presidential Administration," *Harvard Law Review* 114: 2245–2385.

45 James A. Dunn, 2000, "Transportation: Policy-Level Partnerships and Project-Based Partnerships" in *Public-Private Policy Partnerships*, edited by Pauline V. Rosenau, Cambridge, MA: MIT Press: 77–91.

46 Cary Coglianese, 1997, "Assessing Consensus: The Promise and Performance of Negotiated Rulemaking," *Duke Law Journal* 46: 1255–1349.

47 See Jonathan C. Borck, Cary Coglianese, and Jennifer Nash, 2008, "Evaluating the Social Effects of Environmental Leadership Programs," Working Paper of the Regulatory Policy Program, John F. Kennedy School of Government, Cambridge, MA: Harvard University.

48 Theodore Lowi, 1979, *The End of Liberalism*, 2nd edition, New York: Norton.

49 John Graham, 2010, *Bush on the Homefront: Domestic Policy Triumphs and Setbacks*, Bloomington: Indiana University Press.

50 David Lewis, 2008, *The Politics of Presidential Appointments*, Princeton: Princeton University Press.

51 Shirley Anne Warshaw, 2009, *The Co-Presidency of Bush and Cheney*, Stanford, CA: Stanford University Press.

52 Christine Todd Whitman, 2004, *It's My Party Too*, New York: Penguin.

53 William F. West, 2005, "The Institutionalization of Regulatory Review: Organizational Stability and Responsive Competence at OIRA," *Presidential Studies Quarterly* 35(1):76–93.

54 A complete listing and text of the letters can be obtained online from the Office of Management and Budget: www.reginfo.gov/public/jsp/EO/promptLetters.jsp.

55 John Graham, 2010, *Bush on the Homefront: Domestic Policy Triumphs and Setbacks*, Bloomington: Indiana University Press.

56 Christopher McGrory Klyza and David Sousa, 2008, *American Environmental Policy, 1990–2006*, Cambridge, MA: MIT Press.

57 US EPA, Information on Partnership Programs, "List of Programs," /www.epa.gov/partners-index.

58 Marc A. Eisner, 2007, *Governing the Environment*. Boulder, CO: Lynne Rienner.

59 Gary C. Bryner, 2007, "Congress and Clean Air Policy" in *Business and Environmental Policy*, edited by Michael E. Kraft and Sheldon Kamieniecki, Cambridge, MA: MIT Press: 127–151.

60 Michele E. Gilman, 2010, "Presidents, Preemption and the States," *Constitutional Commentary* 26: 339–384.

61 Barry Rabe, 2007, "Environmental Policy and the Bush Era: The Collision Between the Administrative Presidency and State Experimentation," *Publius* 37(3): 413–431.

62 This argument is made by McGarity. See Thomas O. McGarity, 2008, *The Preemption War*, New Haven, CT: Yale University Press.

63 Richard C. Vladeck, "The Emerging Threat of Regulatory Preemption," Working paper, American Constitution Society, January 2008.

64 Melanie M. Marlowe, 2011, "President Obama and Executive Independence" in *The Obama Presidency and the Constitutional Order*, edited by Carol McNamara and Melanie M. Marlowe, Lanham, MD: Rowman & Littlefield: 47–70.

65 John Schwartz, 2009, "From New Administration, Signs of Broader Role for States," *New York Times* (January 30).

66 Office of the Press Secretary, Memorandum, "Preemption," May 20, 2009, www.whitehouse.gov/the_press_office/Presidential-Memorandum-Regarding-Preemption/.

67 Barack Obama, Executive Order 13500—Further Amendments to Executive Order 12859, Establishment of the Domestic Policy Council Online by Gerhard Peters and John T. Woolley, The American Presidency Project, www.presidency.ucsb.edu/node/286200.

68 Justin S. Vaughn and Jose D. Villalobos, 2015, *Czars in the White House*, Ann Arbor: University of Michigan Press.

69 Philip Rucker and Robert Costa, 2017, "Bannon Vows a Daily Fight for 'Deconstruction of the Administrative State,'" *Washington Post* (February 23).

70 Michael Lewis, 2018, *The Fifth Risk*, New York: Norton.

71 Jonathan Mahler, 2018, "All the Right People," *New York Times Magazine* (June 24).

72 Ibid.

73 Karen Orren and Stephen Skowroneck, 2017, *The Policy State*, Cambridge, MA: Harvard University Press.

74 Michael J. Berry, 2016, *The Modern Legislative Veto*, Ann Arbor: University of Michigan Press.

75 Between the enactment of the CRA in March, 1996 and the March, 2001 reversal of the Ergonomics Rule, resolutions of disapproval for twelve other regulations were introduced in Congress. None were successful, and most died in committee. See Morton Rosenberg, "Congressional Review of Agency Rulemaking: An Update and Assessment of the Congressional Review Act after a Decade" (May 8, 2008), Washington, DC: Congressional Research Service Report for Congress, RL30116.

76 This would have been unnecessary under the Clear Skies bill because old and new sources would be treated the same, although the market would create an economic incentive for older power plants to reduce their emissions.

77 Marc A. Eisner, Jeff Worsham and Evan J. Ringquist, 2006, *Contemporary Regulatory Policy*, 2nd edition, Boulder, CO: Lynne Reinner.

78 John Graham, 2010, *Bush on the Homefront: Domestic Policy Triumphs and Setbacks*, Bloomington: Indiana University Press.

79 Even before Clear Skies was announced, EPA enforcement actions dropped significantly in response to cues from the White House in 2001. See Colin Provost, Brian J. Gerber and Mark Pickup, 2009, "Flying Under the Radar? Political Control and Bureaucratic Resistance at the Bush Environmental Protection Agency" in *President George W. Bush's Influence over Bureaucracy and Policy,* edited by Colin Provost and Paul Teske, New York: Palgrave Macmillan: 169–186.

80 Matthew L. Wald, 2012, "Court Blocks EPA Rule on Cross-State Pollution," *New York Times* (August 21).

Bibliography

Anderson, Martin. 1990. *Revolution: The Reagan Legacy*, 2nd edition. Stanford, CA: Hoover Institution.

Balla, Steven J. and John R. Wright. 2001. "Interest Groups, Advisory Committees and Congressional Control of the Bureaucracy." *American Journal of Political Science* 45(4): 799–812.

Berry, Jeffrey and Kent E. Portney. 1995. "Centralizing Regulatory Control and Interest Group Access: The Quayle Council on Competitiveness" in *Interest Group Politics*, 4th edition, edited by Allan J. Cigler and Burdett A. Loomis. Washington, DC: Congressional Quarterly Press.

Berry, Michael J. 2016. *The Modern Legislative Veto*. Ann Arbor: University of Michigan Press.

Borck, Jonathan C., Cary Coglianese, and Jennifer Nash. 2008. "Evaluating the Social Effects of Environmental Leadership Programs." Working Paper of the Regulatory Policy Program, John F. Kennedy School of Government. Cambridge, MA: Harvard University.

Bosso, Christopher. 2005. *Environment Inc.* Lawrence, KS: University of Kansas Press.

Bryner, Gary C. 2007. "Congress and Clean Air Policy" in *Business and Environmental Policy*, edited by Michael E. Kraft and Sheldon Kamieniecki. Cambridge, MA: MIT Press: 127–151.

Coglianese, Cary. 1997. "Assessing Consensus: The Promise and Performance of Negotiated Rulemaking." *Duke Law Journal* 46: 1255–1349.

Duffy, Robert J. 1996. "Divided Government and Institutional Combat: The Case of the Quayle Council on Competitiveness." *Polity* 28(3): 379–399.

Dunn, James A. 2000. "Transportation: Policy-Level Partnerships and Project-Based Partnerships" in *Public-Private Policy Partnerships*, edited by Pauline V. Rosenau. Cambridge, MA: MIT Press: 77–91.

Durant, Robert F. 2007. *The Greening of the U.S. Military*. Washington, DC: Georgetown University Press.

Eads, George C. and Michael Fix. 1984. *Relief or Reform: Reagan's Regulatory Dilemma*. Washington, DC: Urban Institute.

Eisner, Marc A. 2007. *Governing the Environment*. Boulder, CO: Lynne Rienner.

Eisner, Marc A., Jeff Worsham and Evan J. Ringquist. 2006. *Contemporary Regulatory Policy*, 2nd edition. Boulder, CO: Lynne Reinner.

Fiorino, Daniel J. 1995. *Making Public Policy*. Berkeley: University of California Press.

Friedman, Barry D. 1995. *Regulation in the Reagan-Bush Era*. Pittsburgh: University of Pittsburgh Press.

Gilman, Michele E. 2010. "Presidents, Preemption and the States." *Constitutional Commentary* 26: 339–384.

Glassner, Barry. 1999. *The Culture of Fear*. New York: Basic Books.

Graham, John. 2010. *Bush on the Homefront: Domestic Policy Triumphs and Setbacks*. Bloomington: Indiana University Press.

Heclo, Hugh. 1983. "One Executive Branch or Many?" in *Both Ends of the Avenue*, edited by Anthony King. Washington, DC: American Enterprise Institute.

Jones, Charles O. 1977. *An Introduction to the Study of Public Policy*. North Scituate, MA: Duxbury Press.

Kagan, Elena. 2001. "Presidential Administration." *Harvard Law Review* 114: 2245–2385.

Kelley, Christopher S. 2010. "The Unitary Executive and the Clinton Administration" in *The Unitary Executive and the Modern Presidency*, edited by Ryan J. Barilleaux and Christopher S. Kelley. College Station, TX: Texas A&M Press.

Kerwin, Cornelius M. 2003. *Rulemaking: How Government Agencies Write Law and Make Policy*, 3rd edition. Washington, DC: Congressional Quarterly Press.

Kingdon, John. 1995. *Agendas, Alternatives and Public Policies*, 2nd edition. Harper Collins.

Klyza, Christopher McGrory and David Sousa. 2008. *American Environmental Policy, 1990–2006*. Cambridge, MA: MIT Press.

Lewis, David. 2008. *The Politics of Presidential Appointments*. Princeton: Princeton University Press.

Lewis, Michael. 2018. *The Fifth Risk*. New York: Norton.

Light, Paul C. 1999. *The President's Agenda: Domestic Policy Choice from Kennedy to Reagan*, 3rd edition. Baltimore: Johns Hopkins University Press.

Lowi, Theodore. 1979. *The End of Liberalism*, 2nd edition. New York: Norton.

MacKerron, Conrad B. "Has OMB Been Out of Bounds on EPA?." *Chemical Week* (March 22, 1989): 8.

Mahler, Jonathan. 2018. "All the Right People." *New York Times Magazine* (June 24).

Marlowe, Melanie M. 2011. "President Obama and Executive Independence" in *The Obama Presidency and the Constitutional Order*, edited by Carol McNamara and Melanie M. Marlowe. Lanham, MD: Rowman & Littlefield: 47–70.

McGarity, Thomas O. 2008. *The Preemption War*. New Haven, CT: Yale University Press.

McGarity, Thomas O. and Sidney A. Shapiro. 1993. *Workers at Risk: The Failed Promise of the Occupational Safety and Health Administration*. Westport, CT: Praeger.

Meier, Kenneth J. 1985. *Regulation: Politics, Bureaucracy and Economics*. New York: St. Martin's.

Nathan, Richard P. 1983. *The Administrative Presidency*. New York: Wiley.

Obama, Barack H. "Executive Order 13500—Further Amendments to Executive Order 12859, Establishment of the Domestic Policy Council." Online by Gerhard Peters and John T. Woolley, The American Presidency Project: www.presidency.ucsb.edu/node/286200/.

Office of Management and Budget: www.reginfo.gov/public/jsp/EO/promptLetters.jsp.

Office of the Press Secretary. Memorandum, "Preemption." May 20, 2009. www.white house.gov/the_press_office/Presidential-Memorandum-Regarding-Preemption/.

Orren, Karen and Stephen Skowroneck. 2017. *The Policy State*. Cambridge, MA: Harvard University Press.

Pfiffner, James P. 2005. *The Modern Presidency*, 4th edition. Belmont, CA: Thompson Wadsworth.

Provost, Colin, Brian J. Gerber and Mark Pickup. 2009. "Flying Under the Radar? Political Control and Bureaucratic Resistance at the Bush Environmental Protection Agency" in *President George W. Bush's Influence over Bureaucracy and Policy*, edited by Colin Provost and Paul Teske. New York: Palgrave Macmillan: 169–186.

Rabe, Barry. 2007. "Environmental Policy and the Bush Era: The Collision Between the Administrative Presidency and State Experimentation." *Publius* 37(3): 413–431.

Rosenberg, Morton. 2008. "Congressional Review of Agency Rulemaking: An Update and Assessment of the Congressional Review Act after a Decade" (May 8). Washington, DC: *Congressional Research Service Report for Congress*. RL30116.

Rucker, Philip and Robert Costa. 2017. "Bannon Vows a Daily Fight for 'Deconstruction of the Administrative State.'" *Washington Post* (February 23).

Schwartz, John. 2009. "From New Administration, Signs of Broader Role for States." *New York Times* (January 30).

Shafie, David M. 2013. *Eleventh Hour: The Politics of Policy Initiatives in Presidential Transitions*. College Station, TX: Texas A&M University Press.

Shapiro, Stuart. 2007. "Presidents and Process: A Comparison of the Regulatory Process Under the Clinton and Bush (43) Administrations." *Journal of Law and Politics* 23(2):393–418.

Soden, Dennis L. and Brent S. Steel. 1999. "Evaluating the Environmental Presidency" in *The Environmental Presidency*, edited by Dennis L. Soden. Albany: State University of New York Press.

Stone, Deborah. 2002. *Policy Paradox: The Art of Political Decisionmaking*. New York: Norton.

Sunstein, Cass R. 2002. "A New Executive Order for Improving Federal Regulation? Deeper and Wider Cost-Benefit Analysis."

Sutton, Victoria. 2003. "The George W. Bush Administration and the Environment." *Western New England Law Review* 25(1): 221–242.

Tulis, Jeffrey. 1987. *The Rhetorical Presidency*. Austin: University of Texas Press.

US EPA. Information on Partnership Programs, "List of Programs." www.epa.gov/pa rtners-index.

Vaughn, Justin S. and Jose D. Villalobos. 2015. *Czars in the White House*. Ann Arbor: University of Michigan Press.

Vig, Norman J. 2006. "Presidential Leadership and the Environment" in *Environmental Policy: New Directions for the Twenty-First Century*, 6th edition, edited by Norman J. Vig and Michael E. Kraft. Washington, DC: Congressional Quarterly Press. 100–123.

Vladeck, Richard C. "The Emerging Threat of Regulatory Preemption." Working paper. American Constitution Society, January 2008.

Wald, Matthew L. 2012. "Court Blocks EPA Rule on Cross-State Pollution" *New York Times* (August 21).

Warshaw, Shirley Anne. 1994. "The Carter Experience With Cabinet Government" in *The Presidency and Domestic Policies of Jimmy Carter*, edited by Herbert D. Rosenbaum and Alexej Ugrinsky. Westport, CT: Greenwood.

Warshaw, Shirley Anne. 2009. *The Co-Presidency of Bush and Cheney*. Stanford, CA: Stanford University Press.

Waxman, Henry. 2008. "Opening Statement of House Committee on Oversight and Government Reform, Business Meeting Regarding Contempt Resolution" (June 20). One Hundred Tenth Congress.

West, William F. 2005. "The Institutionalization of Regulatory Review: Organizational Stability and Responsive Competence at OIRA." *Presidential Studies Quarterly* 35(1): 76–93.

Whitman, Christine Todd. 2004. *It's My Party Too*. New York: Penguin.

3

MANAGING THE COMMONS

President Trump's Interior Secretary brought some dramatic flair to his first day on the job. Announcing there was a new sheriff in town, Ryan Zinke arrived for work on horseback, escorted by a mounted unit of national park police. After Zinke and his horse Tonto rode across the National Mall to the Interior Department, a special flag was raised above the building to announce his presence. Then in his first official act as secretary, he signed an order rescinding President Obama's ban on lead ammunition on federal lands.[1] He then turned his sights on other wildlife protection policies, including several rules intended to implement Endangered Species Act (ESA). Rules protecting migratory birds, restrictions on hunting grizzly bears, and protections for marine mammals and sea turtles were all suspended.[2] Zinke would resign less than two years later, amid mounting personal scandals and 11 criminal investigations, but not before accomplishing several of Trump's goals for the Interior Department.

The ESA was long despised by resource-extractive industries and property rights advocates. Zinke's deputy and successor was David Bernhardt, an agribusiness and oil lobbyist who had sued the Interior Department and written an op-ed in the *Washington Post* calling the 45-year old law a "regulatory burden," and proposing that Congress strip legal protections from threatened species.[3] The law has remained popular with the public, so legislative repeal has never been feasible. Anti-ESA sentiment was high in the Conservative Congress that arrived in Washington in 1995, but the most Congress could do was tame the law with a rider that temporarily prohibited the Fish and Wildlife Service (FWS) from listing new species. Repeal was no more of an option during the years when George W. Bush was president and Republican majorities held Congress. Rather, a quiet campaign to weaken the ESA was waged through administrative channels. A 2008 rule waived the ESA independent scientific review requirement.[4] This gave agencies the discretion to decide on their own whether there was any potential harm to an endangered species, effectively making the reviews optional. The rule change, giving agencies the discretion to opt out of the ESA's independent review requirement, was characterized in a *New York Times* editorial as an attempt "to rewrite the law, albeit through regulatory means."[5]

There had been other efforts to curtail ESA implementation during the previous eight years. In 2004, the Union of Concerned Scientists claimed that Bush appointees at the FWS discouraged challenges to federal projects and interfered with scientific reviews. According to the report, 20 percent of the agency's scientists claimed that they were pressured to alter their reports to downplay threats to endangered species.[6] A 2005 agency directive authorized FWS staff to use any information from agency files to refute petitions filed with the agency, but prohibited its use in support of those petitions.[7] In several other instances, the administration was alleged to have used the rulemaking process to intervene on behalf of an interest group. For instance, when Congress debated reauthorization of the Magnuson-Stevens Fisheries Act, the fishing industry lobbied unsuccessfully to relax the environmental review requirements used to set commercial catch limits. Subsequently, the administration proposed a rule that would have given regional fisheries councils the power to reject management plans prepared by the National Marine Fisheries Service, but this was withdrawn after Obama's election.[8] For the most part, these efforts paid off. On average, eight new endangered species were listed annually when Bush was president, down from 58 per year during his father's presidency and 63 annually during the Clinton years.[9]

For Obama to rescind the 2008 Bush rule, it would have meant initiating a new rulemaking process, or spending political capital to persuade Congress to strengthen the law. Instead, he issued a directive to the FWS declaring the independent reviews were essential tools in the enforcement of the ESA and ordered them to continue. This had the effect of reversing the Bush-era rule because he trusted agency personnel to follow his policy directive because they agreed with its goals. At least for the time being, leadership from the top provided clarity and continuity with regard to endangered species. Obama's administrative strategy strengthened the ESA in the short term, but his changes were easily undone by the next administration. Secretary Zinke went further by proposing rule changes that would make it easier for pipelines and other construction projects to gain approval. The new rules, which took effect in 2019, rolled back protections for species only listed as "threatened," and instructed agencies to begin considering economic factors in deciding whether a species should be protected, and to disregard the effects of climate change.[10]

This chapter examines the expansive role of executive leadership over public land management following the era of Congressional dominance in environmental policymaking. In the 1960s and 1970s, federal officials reexamined their practices amid mounting evidence that "efficient" use of natural resources was affecting the ecology of public lands. A mobilized environmental community in the postwar era awakened preservationist sympathies in the mass public, and Congress responded with the Wilderness Act of 1964, withdrawing sections of federal land and placing them off limits to development permanently. Postwar prosperity also fueled two incongruous trends: demand for new housing increased the rate of clear-cutting for lumber, and increasing leisure time and the proliferation of automobiles stimulated demand for public areas for recreation. By enacting the Federal Land Policy and Management Act (FLPMA) in 1976, Congress sought to resolve this tension by requiring agencies to balance development with recreational uses. Despite these major statutes, wilderness preservation and multiple use continued to be the dominant sources of conflict over public land policy.[11]

The Bureaucratic Landscape

Of four agencies that share responsibility for managing federal lands, the FWS is the smallest. Two more located within the Department of Interior's hierarchy are the Bureau of Land Management (BLM) and the National Park Service (NPS). They, along with the US Forest Service (a division of the Department of Agriculture) are charged with carrying out policies enacted by Congress. Because of the number of statutes and implementing agencies with conflicting and ambiguous mandates, the presidency is well-positioned to resolve conflicts and set priorities for this complex of bureaucratic structures. The White House plays a central role in natural resource policymaking largely because of the resurgent Administrative Presidency of Ronald Reagan. Although the popular notion of a "revolution" during the Reagan era was hyperbole, James Watt's Interior Department represented an ambitious attempt to reorganize the federal bureaucracy. Through administrative reforms, political appointees and presidential directives, Reagan redirected policy away from preservation and toward more aggressive development of natural resources.

The FWS was one agency that experienced budget cuts and multiple reorganizations during the Reagan years. These administrative actions further hampered an organization that lacked adequate resources to fulfill its statutory mandates. One of several bureaus, services and offices housed within the Interior Department, the FWS lacked the organizational clout of its larger sibling, the National Park Service (NPS). Ever since its origin as the Bureau of Fisheries in 1903, the FWS had been relocated from the Department of Commerce and Labor to the Interior Department, merged with the U.S. Biological Survey and then split up, and undergone several other changes in mission and structure. As Richard Tobin remarks, "these frequent reorganizations reflect poorly on the Service's political support and administrative strength."[12] Depriving the agency of resources had significant implications for species preservation because of its role in implementing the ESA, a responsibility it shares with its step-cousin at the Department of Commerce, the National Marine Fisheries Service.

The Reagan White House placed limitations on the agency's capacity to implement the ESA without devoting resources to statutory reform. As discussed in Chapter 2, Reagan's strategy toward ENR policy was to use administrative tools to achieve his goals at the implementation stage. This provoked a confrontation with environmental groups, who pursued litigation against the administration and pressured Congress to strengthen laws and restrict agency discretion.

By the early 1990s, federal land policy was hopelessly deadlocked. Congress had passed historic statutes to bring balance to federal land management practices, but offered little guidance in resolving the continuing controversies. Federal laws required lands to be managed for multiple uses, and land management agencies were caught in the middle of conflicting demands. Presidential initiative made it possible to break the stalemate. When Bill Clinton arrived in this new political landscape, he experienced some success working with Congress on popular bills to preserve land in California and Colorado, but not with bills to reform the far more contentious areas of mining and grazing policy. Clinton's significant achievements were his administrative initiatives, especially when he threw his support behind the ideas of agency officials, after earning their loyalty and trust.

From Owls to Ecosystems

For much of the century after its founding, the Forest Service enjoyed an outstanding reputation for competence and cultivating public trust because of its ability to interpret statutory mandates and manage multiple goals.[13] The organization marched to a single drummer, and there was agreement that harvesting timber was the centerpiece of its mission. This meant clearcutting and replanting entire forests, in a practice known as sustainable yield. By the 1970s, conservation scientists were questioning the viability of this approach, and the Forest Service began to study ecosystem preservation as an alternative. Furthermore, preservation activists scored a victory by calling the statutory basis for clearcutting into question.[14] The Forest Service experienced a cultural sea change as a new generation of foresters entered the ranks. Reflecting the prevailing view that clearcutting for maximum yield was unsustainable, Forest Service Chief Dale Robertson in 1992 declared that ecosystem management should be the guiding principle for agency decisions.[15]

The spotted owl controversy represents a case in point. Wildlife biologists had been studying the northern spotted owl, whose population was almost completely in old-growth forests of Oregon, Washington, and Northern California. Along with local members of the Audubon society, they began to sound the alarm about its rapidly-dwindling habitat in the 1970s. Early on, federal land management officials considered setting aside land to help the species recover, but decided against it because of a lack of certainty over the number of acres it required, and fears about the precedent restricting land use for a single species.[16]

When the ESA was enacted in 1973, the spotted owl received consideration as a potential endangered species. After the FLPMA became law, a task force representing state and federal land management agencies convened, and hoping to avoid ESA listing, issued a plan to protect 400 nesting pairs (with 300 acres per pair) on Forest Service, BLM and state land in Oregon. After a backlash from the timber industry and logging communities, the agencies agreed to revisit the plan after five years, but tension mounted in the ensuing years. The timber industry predicted job losses and pressed for more harvesting, and environmentalists argued for more preservation as scientific research on old growth forests suggested that more than three times as much intact forest was needed to maintain a viable habitat. After the FWS denied a petition to list the spotted owl as endangered, environmentalists brought a lawsuit to the U.S. District Court in Washington State to block timber sales.

Congressional involvement in the late 1980s and early 1990s failed to diffuse the controversy. Rather, a series of legislative riders on behalf of environmentalists and timber companies compounded the tension. One judicial ruling, in particular, pushed the controversy past the boiling point. In 1990, the Ninth Circuit Court in California sided with environmentalists and struck down a legislative rider that forced timber sales to continue. When Interior Secretary Manuel Lujan of the George H.W. Bush Administration proposed a 20 percent reduction in timber sales to comply with the decision, the proposal was rebuffed in Congress. Instead, Democrats in Congress pressed for a one-third reduction in timber sales, but were unable to bring their proposal to a vote in the senate. President Bush did little to quell the controversy by dismissing environmentalists as people who "care more about birds than jobs," as he waged a losing campaign for reelection against challenger Bill Clinton.

The controversy was severe enough that Clinton became personally involved. One of his first acts as president was to hold a forest summit in Oregon. Clinton established a Forest Ecosystem Management Team, tasked with developing a plan that would meet the requirements of court orders and satisfy the logging communities, timber interests and environmentalists. The result was the Northwest Forest Plan, issued July 1, 1993. The plan set aside large sections of forest to protect entire watersheds, while still permitting the harvesting of up to 1.2 billion board-feet from old growth forests, constituting a substantial drop from the volume permitted in the late 1980s. The plan also permitted salvage logging in old growth areas, and provided funds for job training and conservation programs that would employ potentially displaced workers.[17]

The Forest Summit and Northwest Forest Plan broke the impasse and inspired collaboration in other communities. In one of the most successful early examples of collaborative management, activists in a Northern California community formed the Quincy Library Group, which overcame similar conflicts over spotted owl habitat to develop its own forest plan. The collaboration succeeded because it was seen as a better alternative to having a plan imposed upon the community from the outside. Members also stressed values they held in common, namely a commitment to preserving their community lifestyle and local control. Moreover, neither side identified with urban environmentalists, who they distrusted. The group credited Clinton's involvement with the Forest Summit, particularly his appeal to abandon the courtroom and search for "win-win" solutions.[18]

The Northwest Forest Management Plan was soon undermined by timber interests and their allies in the 104th Congress. As Chapter 2 argued, a role-reversal occurred after the 1994 election, the new Republican majority pursued legislation to rein in the regulatory state, and the White House became its principal defender. Much of this antiregulatory activism took the form of "unorthodox" lawmaking, particularly legislative riders.[19] In 1995, the House of Representatives used this tactic to triple the volume of "salvage" timber. The rider effectively raised the timber quota because it authorized the Forest Service to harvest and sell "associated green trees" in addition to the diseased and dying timber normally classified as salvage timber. The bill's definition of salvage timber was so broad, New Jersey Senator Bill Bradley claimed, that it could justify "harvesting any tree made of wood."

The salvage provision was one of several riders attached to the 1995 Rescissions Act, a spending bill that included assistance for victims of the Oklahoma City Bombing. Citing the riders, Clinton vetoed the bill, but later signed another version that included the salvage rider. Clinton expressed confidence that his administration could implement the provision in an ecologically sound manner, but the language gave the Forest Service little flexibility, requiring timber sales and logging even within old-growth areas. Interior Secretary Bruce Babbitt later conceded that "unprepared White House negotiators gave away the store when they agreed to a so-called salvage rider."[20]

Roadless Areas

Clinton was better prepared two years later when Congress sent him an appropriations bill with a rider that prohibited the Forest Service from completing planned scientific reviews of impacts on endangered species before issuing new forest management plans. Even though he signed the bill, he attached a signing statement directing the USDA to proceed with the reviews. The 1,400-word statement also scolded Congress for micromanaging forest policy and undermining the role of science:

> The Congress also continues to interfere with the Administration's efforts to promote ecosystem management and a greater understanding of the natural resource management issues affecting areas like the interior Columbia River Basin—an area characterized by forest health, watershed, and endangered species problems....This action may benefit a few special interests, but it injures both the environment and the economy.

The mention of ecosystem management was a welcome surprise for Mike Dombeck, the new director of the Forest Service. Dombeck was the latest in a series of leaders of the Forest Service with a background in ecology, committed to preserving intact ecosystems as a method to maintaining forest health. He was also struck by the growing discrepancy between the Forest Service's mandates and its resources, and he had been eyeing the agency's expenditures on road construction.

Dombeck had been working to convince lawmakers that the taxpayers were losing money because of timber quotas. There had been a contentious debate in Congress over cutting the Forest Service's road budget as a way to save tax dollars. For most of its century of existence, the Forest Service financed road construction in national forests with revenue from timber sales. Much of the remaining timber stock in the 1990s was on remote terrain, which required longer roads on steep grades to harvest. Moreover, the quality of the remaining timber and the nature of the terrain reduced the yield. The operation, which included maintenance on over 340,000 miles of logging roads, had an $8.4 billion annual shortfall. In 1997, the House of Representatives came within a single vote of passing a bill to terminate the agency's road-building budget completely.[21]

When the salvage logging rider expired, Dombeck suggested to Agriculture Secretary Dan Glickman that a new review of roadless areas should be conducted for a new transportation policy for the National Forest System.[22] To his surprise, Glickman responded by asking for a proposal. Dombeck then proposed a temporary moratorium on road building to conduct the review, which received the president's blessing and fast-track authorization. The review was completed in 18 months.

In an example of taking ownership of his administration's initiatives, Clinton chose a scenic autumn vista in Virginia's Shenandoah Valley to announce that his administration would develop a comprehensive policy to protect the National Forest System's remaining wild areas. Several months later, the USDA proposed its new transportation policy and conservation plan for roadless areas, which included a ban on new road construction and replacement in 49 million inventoried areas, allowing timber harvesting only for reduction of fire risk, habitat protection and other stewardship purposes. During the 60-day comment period and 190 public meetings, the Forest Service received over 1.1 million comments, 95 percent in support of the proposal.[23]

The proposed Roadless Area Conservation Policy dealt gingerly with the question of the mostly road-free Tongass National Forest, which Alaska state officials hoped to develop. The prevailing assumption was that including the Tongass could provoke a backlash that would imperil Vice President Gore's presidential campaign. With the presidential election in the balance, the Forest Service recommended a compromise allowing road construction in the Tongass continue until 2004, when the number of protected acres would increase to 58.5 million.[24] Once the Supreme Court issued its decision in Bush v. Gore, resolving the question of presidential succession, Clinton surprised everyone by announcing a directive to include the Tongass immediately. "This will include protection for the last great temperate rainforest in America," Clinton announced in a speech at the National Arboretum in January, 2001. Getting in front of the issue in characteristic fashion, Clinton declared he was making the roadbuilding moratorium permanent, as well as ending commercial logging and mineral leasing in all 58.5 million acres of roadless area.[25]

The roadless rule appeared in the *Federal Register*, January 12, 2001, and was promptly frozen in a 60-day moratorium by incoming Chief of Staff Andrew Card, along with other pending regulations.[26] Since it had been less than 60 days, the incoming Bush Administration could legally delay the rule from taking effect. In order to overturn this and other policy initiatives left behind by the Clinton Administration, Bush used a dual strategy of waiting for interest groups to challenge the rule in court, while developing a more industry-friendly alternative policy.[27]

The Bush Administration made no effort to defend the roadless rule against pending litigation in Idaho and Wyoming. In response to a challenge by the state of Idaho and the Boise Cascade Company, the Justice Department persuaded the judge to delay the implementation date another 60 days so that the new administration could study the rule. Following the second delay, the administration determined that the scoping phase was flawed and declared its intent to amend the rule.[28] Nonetheless, the rule was struck down for this very reason in U.S. District court in Idaho, resulting in a nationwide injunction. The rule was also vacated in U.S. District Court in Wyoming, by a judge who determined that the rule usurped the power of Congress to declare wilderness areas. Once the courts did the work of striking the rule, the Bush Administration proposed a new rule in 2004 that allowed states to propose their own management plans for roadless areas. The process dragged on as the USDA was overwhelmed with 1.8 million comments from environmentalists, hunters and anglers, mostly in opposition, before the rule was completed in 2005. By then, the White House was no longer directing the process, and the dueling decisions worked their way through the federal courts.[29]

Healthy Forests

The Bush Administration already signaled a new direction for forest policy with its refusal to defend Clinton's roadless rule. In 2002, Bush took aim at the Northwest Forest Management Plan with his "Healthy Forests" initiative, which directed the Forest Service and BLM to develop new rules to reverse features of the plan that were seen as obstacles to logging. One was the survey-and-manage rule, requiring agencies to take measures to protect species that depend on old-growth habitat. The other was a requirement to protect salmon by maintaining a buffer zone around streams. The new rules were promulgated in 2004 amid protests from scientists within both agencies, and from outside of government, and after a series of legal challenges the Ninth Circuit Court ordered that the survey-and-manage rule be reinstated.

Because Bush was unable to implement his entire forest agenda administratively, he proposed legislation to allow additional timber harvesting through language similar to the expired salvage rider. In the name of fire prevention, the legislation allowed "thinning" of densely forested areas as a way to reduce fuel. The bill also increased harvesting of larger, more economically valuable trees as a way to finance the harvest of smaller trees. After an early version was blocked by a filibuster, a scaled-back version passed the Senate as the Healthy Forests Restoration Act of 2003. The stewardship contracting provision passed as a rider. The bill also limited judicial review of timber sales and restricted administrative appeals. Thus, the policy was a "perfect vehicle for increased administration's discretion."[30]

The Bush Administration's experience with forest policy illustrates the limits of administrative policymaking as well as its power. The Clinton Administration left behind a set of management practices that were grounded in science, justified under his statutory authority, and had the support of the civil service. The Bush strategy of administrative reversal resulted in policies that were less resilient. His initiatives mostly succeeded in delaying Clinton-era policies for short-term increases in timber production. The forest policy inherited by Obama was a confusing tangle of conflicting court rulings and rules. The new Agriculture Secretary, Tom Vilsack, had little choice but to freeze new road construction as the administration waited for the federal courts to hear challenges and sort the policy out.[31]

The roadless rule ultimately survived, but ecosystem management as a federal policy did not fare as well. The Bush Administration saw ecosystem management as an obstacle to resource development. Vice President Cheney's national energy task force recommended increased oil and gas leasing on federal land, and the new administration modified the policy in the service of its goals. James Skillen argues the Bush Administration rejected the substantive model of ecosystem management, while using the procedural aspects to its advantage.[32] The administration championed the collaborative planning already in use by federal land management agencies, but local participants and economic interests would be given priority. Bush formalized the new policy of "cooperative conservation" in a 2004 executive order.

Ecosystem Management and the National Parks

Regarding the national parks, the Clinton Administration found statutory authority to promote ecosystem integrity in the language of the NPS Organic Act of 1916. The law required the NPS to balance preservation and use, as well as the discretion to make policy whenever there is conflict between the two. Park Service personnel argued that the preservation mandate had been neglected, and they found the administration receptive to their message. Ecosystem preservation continued to be a cause for administrative actions.

After years of inaction, the NPS in 2000 responded to mounting evidence of environmental degradation by prohibiting snowmobiles in the Yellowstone region. The Record of Decision cited research on air pollution and the impact on wildlife that supported restrictions on snowmobile use.[33] Park service personnel within Yellowstone experienced these effects directly every winter, suffering the same noise-induced stress and respiratory problems as the wildlife. The administration exceeded their expectations by endorsing an outright ban, which was justified under the Organic Act's preservation mandate.[34] All private snowmobiles were to be excluded from Yellowstone, as well as the Grand Teton National Park and John D. Rockefeller Memorial Wilderness. In response, off-road vehicle manufacturers claimed that communities surrounding Yellowstone could lose up to 1,000 jobs and $100 million annually in tourism revenue, and Wyoming Republican Senator Craig Thomas threatened legislation to reverse the ban.[35]

The incoming Bush Administration, already on record as opposing the ban, declined to defend the rule in court against the inevitable challenges. The opening salvo came from the Snowmobile Manufacturers Association, which filed suit in U.S. District Court in Wyoming as soon as the final rule was listed. Using the same approach as with the roadless rule, the administration's tactic was to settle the lawsuit and commit to a new rulemaking as part of the agreement.[36] When the administration proposed its replacement rule in 2003, it was challenged by the Greater Yellowstone Coalition. This time, the rule was struck down because it singled out snowmobile operation as an authorized use, thus it failed to balance preservation mandate along with use, as required by the Organic Act. Moreover, the judge ruled that the rule violated the arbitrary and capricious standard, since the NPS neglected to consider evidence of the ecological impact of snowmobiles that its own scientists provided in developing the previous rule. The administration settled again, leading ultimately to a compromise in the 2007 Winter Use Plan which allowed limited snowmobile use in the parks, but only by licensed tour operators.[37]

Wilderness Preservation

One area where Congress took the initiative during the environmental era of the 1960s and 70s is wilderness preservation. Millions of acres were withdrawn from future development after the Wilderness Act was passed in 1964. Over the next four decades, large and small parcels of federal land were studied as suitable candidates for wilderness, and withdrawn legislatively by every Congress. Figure 3.1 illustrates the number of acres each Congress preserved since 1993–94. In that session, Congress passed the California Desert Protection Act, preserving more land than any session except 1965, the first year of the Wilderness Act, and 1980, when the Arctic National Wildlife Refuge was created. Although the amount has varied over the years, the average number of acres preserved through legislation declined after 1994. The 112th Congress was the first one that failed to preserve any federal land as wilderness. This was not for a lack of interest. In that legislative session, 30 conservation bills stalled that would have withdrawn over 5 million acres, even though they were sponsored by home state senators and representatives.

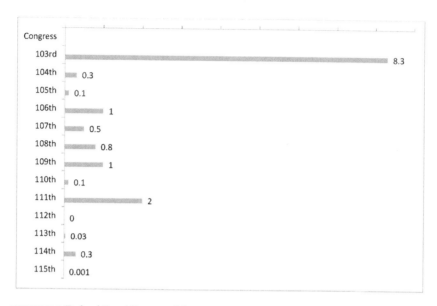

Congress

103rd	8.3
104th	0.3
105th	0.1
106th	1
107th	0.5
108th	0.8
109th	1
110th	0.1
111th	2
112th	0
113th	0.03
114th	0.3
115th	0.001

FIGURE 3.1 Federal Land Preserved by Congress, 1993–2018 (million acres)

So extensive was the aversion of the 112th Congress to consider wilderness designation that it blocked an administrative initiative that would have accomplished many of the same goals. At the end of 2010, Interior Secretary Salazar signed Secretarial Order 3310, which directed the BLM to study remote undeveloped land with wilderness characteristics for designation as "Wild Lands." Under the directive, suitable lands would have received the designation after public hearings and a comment process, and then maintained to preserve their value as wilderness. There was an immediate outcry from Congressional Republicans representing western states and various interest groups, including the National Beef Cattleman's Association, which slammed the order as just an attempt to create de facto wilderness. Taking advantage of the ongoing battle over the federal debt ceiling and spending authorizations, opponents added a rider to a continuing resolution that prohibited the Interior Department from using any funds for the purposes of Wild Lands study or designation. On June 1, 2011, Salazar withdrew the order.[38]

National Monuments: Preservation by Proclamation

Two months before the 1996 presidential election, a photo appeared in the nation's newspapers that featured Bill Clinton in front of a spectacular backdrop, signing an order to designate a new national monument. The picture was striking because of the incongruity of the scene. Outdoors, seated at a desk perched on the edge of a canyon, he created a new 1.7 million-acre National Monument in

Southern Utah with the stroke of a pen. Unlike a signing ceremony for legislation, where the president is flanked by leaders of Congress and others who share credit, the only other soul in the empty landscape was Vice President Albert Gore. The photo implied that the landscape was part of the newly designated monument, but the text revealed a crucial detail: the signing ceremony was actually held at a more politically hospitable venue, the South Rim of the Grand Canyon in neighboring Arizona.[39]

The proclamation which created Grand Staircase-Escalante National Monument followed years of gridlock and stalemate. The FLPMA of 1976 directed the BLM to conduct and inventory of roadless areas and assess their potential for designation as wilderness. The BLM identified suitable areas as wilderness study areas (WSAs) and was directed to preserve them until Congress could make a decision about which areas to preserve as wilderness. Once the inventory of Southern Utah was complete, environmentalists challenged it immediately. In their appeal to the BLM, the coalition argued that much of the inventory was conducted by helicopter, without validating their observations through fieldwork on the ground, and without taking ecosystem management into account. In 1985, the coalition organized volunteers to conduct their own inventory of the WSA. Five years later, they presented their own proposal for the Red Rock Wilderness, an area with much larger boundaries than recommended by the BLM.

By 1995, Congress was no closer to following through on its promised wilderness designation. The Southern Utah Wilderness Alliance was formed to advocate the preservation of the Red Rock Wilderness, an area nearly 6 million acres in size, constituting over one-fourth of all BLM-managed land in the state of Utah. This was the basis for one of two competing proposals for wilderness legislation in the 104th Congress. The Alliance lobbied in support of the "Citizens' Bill," which would have designated wilderness status upon 5.7 million acres of roadless area, including Grand Staircase, Escalante Canyon, Cedar Mesa, and areas surrounding Canyonlands, Zion, and Glen and White Canyons. A competing proposal was favored by Utah's state legislature and Congressional delegation. Introduced in the House by Rep. James Hansen and Senate by Sen. Orrin Hatch, this version would have designated only 1.8 million acres. In addition to protecting a much smaller area, the Utah delegation's proposal also contained a "hard release" provision, to preclude any further wilderness designations in the state.

In the summer of 1995, there was intense debate over the fate of the WSA. Wilderness supporters showed up in large numbers at two public meetings in Salt Lake City, and newspapers throughout the West ran editorials in support of the Citizens' bill. However, the after the Republican takeover of both houses of Congress, the oil, coal, and mining industries enjoyed more influence than they had in many years, and the environmental groups suffered a defeat when their preferred bill died in committee. The Utah delegation's proposal became the only viable legislative solution to end the controversy. The House and Senate were considering identical versions of the bill, which had the support of the mining and coal industries. However, the proposal's hard release provision ensured that the remaining 3.9 million roadless acres not included in the wilderness would be open for multiple use development. At this point, the environmental coalition shifted to a different venue to pursue its goal.

Wilderness advocates took advantage of existing networks, including personal with the White House. Their point of contact was presidential senior adviser George Stephanopoulos. The executive director of the Southern Utah Wilderness Alliance, Mike Matz, knew Stephanopoulos from the 1980s, when Matz worked for the Sierra Club and Stephanopoulos was a staffer for Representative Richard Gephardt. Following the Exxon Valdez disaster, the two collaborated on comprehensive oil spill legislation that Gephardt sponsored in the House. In 1995, Peter Hoagland, a former three-term representative from Nebraska, was working at a Washington law firm that represented the Alliance on a pro-bono basis. Hoagland reached out to Stephanopoulos, requesting a meeting express his concerns about the Republican bill.[40]

In September, the Alliance presented Stephanopoulos with a detailed political analysis, urging the president to oppose the Republican bill. Touting the results of a pair of focus groups in California and Massachusetts, the Alliance stressed that a veto would "reinforce the theme that Republicans are out of step with mainstream Americans and in bed with special interest groups" (SUWA, 1995). The memo further predicted a veto would be popular in key states on the East and West coasts as he prepared to run for reelection:

> *You have nothing to lose in Utah.* Only 26 percent of Utahns support Hansen and Hatch's plan. Even the Salt Lake Tribune called it a "disappointment." Clinton can't win in Utah. *You have a lot to gain regionally, with conservationists an in California.* This issue resonates with New West voters – Clinton's core vote in 1992. *Opposing this is on message.* Hansen and Hatch's bill (H.R. 1744 and S. 884) is another giveaway to the special interests. Republicans raiding the cookie jar for their corporate buddies. It puts big mining corporations first and the people of Utah and our country last (SUWA, 1995; emphasis in original).

Soon after that, Clinton declared his opposition. The legislative debate remained in stalemate, as Republicans were unwilling to consider the Citizens bill, but reluctant to pursue their own bill in the face of a veto threat. Acting on a recommendation from Babbitt and the Interior Department, Clinton invoked the Antiquities Act for the largest preservation declaration since the presidency of Theodore Roosevelt.

As the environmental coalition predicted, the election-year monument designation was popular with the public. It also angered politicians and rural communities in the region, who had hoped to mine rich seams of coal beneath the scenic landscape. That November, Utah's lone Democratic representative in Congress was defeated in an election that gave Clinton a second term and nominal gains for Democrats nationally.[41] Fearing Clinton had more plans to use the power he recently re-discovered, Congressional Republicans threatened to rewrite the Antiquities Act.[42] Waiting until the final year of his presidency, he created seven other national monuments by Election Day, 2000, and 11 more before stepping down in January, 2001. Table 3.1 presents the number of acres Clinton and his successors preserved through administrative actions. These last-minute monuments were Clinton's major environmental legacy.[43]

President George W. Bush created a distinct legacy of preservation by acting to protect large swaths of the Pacific Ocean from development. Administrative withdrawals of federal land were deeply unpopular with Republican constituencies and Congressional Republicans, so Bush made no national monument designations to preserve federal lands. However, late in his presidency, Bush granted national monument status to more than 200 million acres of marine ecosystem. The four national marine monuments vastly exceeded the volume of land preserved administratively by any of his predecessors.[44] By comparison, the land area of national monuments created by Clinton was just over 5 million acres.

Congress became more active during the first two years of Obama's presidency, passing legislation to withdraw more than 2 million acres from economic development (see Figure 3.1). As Obama prepared to run for reelection, there were murmurs that the administration was considering a unilateral action like Clinton's election-year monument designation after a BLM document was leaked to the Utah Congressional delegation.[45] The 2010 draft memo referred to the "Treasured Landscapes" Initiative, which Interior Secretary Ken Salazar advocated to consolidate the bureau's fragmented lands as National Conservation Areas. The paper proposed that BLM work with the administration and Congress to identify which of 14 parcels of public land might be suitable to qualify as conservation areas, which the bureau could preserve and better manage as complete ecosystems:

> Should the legislative process not prove fruitful, or if a nationally significant or cultural land resource were to come under threat of imminent harm,

TABLE 3.1 Federal Land Preserved by Administrative Action, 1993–2016

	Land Monuments (acres)	Marine Monuments (square miles)
Clinton	5,418,153	0
Bush	0	341,865
Obama	5,687,665	855,995

Note: No new National Monuments were created between 1978 and 1996.

BLM would recommend that the Administration consider using the Antiquities Act to designate new National Monuments by Presidential Proclamation.[46]

Not surprisingly, passages like the one above raised suspicions among western and Republican politicians. There were lingering bad feelings from the 20 times that Clinton invoked the Antiquities Act to create or expand national monuments in the final year of his presidency. The Senate rejected a Republican amendment that would have blocked the Administration from declaring any of the 14 parcels as monuments, but its author, Senator Jim DeMint, spoke for many westerners who were still sore about Clinton's midnight monuments:

> A secret administration memo has surfaced revealing plans for the federal government to seize more than 10 million acres from Montana to New Mexico, halting job-creating activities like ranching, forestry, mining and energy development.... President Obama could enact the plans in the memo with just the stroke of a pen, without any input from the communities affected by it.[47]

In response, the National Beef Cattleman's Association called for legislation to amend the Antiquities Act to require Congressional approval for presidential monument declarations.

If Obama intended to follow Clinton's example by creating new monuments before reelection, he did so with caution. Waiting until one year before the election, Obama declared monuments with historic as well as ecological significance. The first was in the battleground state of Virginia: Fort Monroe National Monument, an island at the mouth of the Chesapeake Bay that had served as a Union outpost during the Civil War and a haven for escaped slaves. The action was criticized by Republican leaders such as House Speaker John Boehner, but praised by Virginia's Republican Governor. As his reelection drew nearer, he asserted this power more freely. In April 2012, he created Fort Ord National Monument by combining over 7,000 acres of land from the former army base with another 7,000 acres already managed by the BLM, preserving a diverse and sensitive landscape adjacent to California's Monterey Bay. In September, he issued another proclamation to protect 4,726 acres in Colorado as Chimney Rock National Monument and created Cesar Chavez National Monument in Delano, California. In all four cases, Obama declared that the land amounted to "the smallest area compatible with the proper care and management of the objects to be protected." Obama stepped up his unilateral preservation actions in his second term. After the 112th Congress adjourned without voting on any of the preservation bills introduced during the previous two years, Obama acted unilaterally to preserve five of the designated sites as national monuments.[48] After Senator Harry Reid tried and failed to enact wilderness legislation to protect unique desert terrain in his home state of Nevada, preservationists shifted their focus to the White House. In response, Obama acted unilaterally to create the 704,000-acre Basin and Range National Monument in 2015.[49]

The Battle for Bears Ears

President Barack Obama used his authority under the Antiquities Act to create or expand 29 national monuments, mostly in the final months of his presidency. At 1.35 million acres, Bears Ears in Southeastern Utah was not the largest (only a 1.6 million-acre monument in California's Mojave Desert was larger) but it was the one that would trigger the most consequential backlash. In an echo of the Grand Staircase-Escalante controversy of 20 years earlier, a coalition of tribes and environmentalists petitioned the White House to use the Antiquities Act to large expanse of proposed wilderness in San Juan County, Southeastern Utah. Bears Ears was a 1.9 million-acre conservation area, named for a pair of 8,700-foot buttes, which rise above the Colorado Plateau. The protected area includes other unique geographic features such as Cedar Mesa, as well as Native American archeological and historic sites which the Navajo, Ute Mountain, Hopi, Zuni, and Ute nations sought to protect from vandalism.

The canyons, mountains, and mesas of Bears Ears are not a static museum of antiquities. They remain vital as a place of subsistence, spirituality, healing, and contemplation for the Colorado Plateau's native peoples. Families, elders, and traditional practitioners come quietly to this land to hunt and to gather herbs and firewood, to visit ancestral sites, to perform ceremonies—"to receive nourishment for our spiritual and psychological wellness."[50] The idea to designate Bears Ears a national monument was first floated to President Franklin D. Roosevelt by Interior Secretary Harold Ickes in 1936, but its status remained unresolved long after the public land inventory.

One major difference between the Bears Ears coalition and earlier preservation movements was that the primary advocates were Native Americans tribes. The Inter-Tribal Coalition representing the various native nations formed a nonprofit group, Diné Bikéyah, which means "Peoples Sacred Lands" in the Navajo language. In 2012, the Navajo Nation and San Juan County commissioners entered into a joint planning process, which resulted in an agreement to pursue legislation to preserve all 1.9 million acres. Their proposal failed to gain traction with the state's Congressional delegation, which was concerned about the consequences for ranching, mining and energy interests.

The status of Bears Ears became a cause for public concern again in 2015, when the Utah Legislature drafted a bill proclaiming an "Energy Zone," around Cedar Mesa. The resolution declared that grazing and mining were the "highest and best uses" of the land. In the face of public backlash, the sponsors added language to the resolution declaring that grazing, mineral, and energy development would be conducted in ways that "protect and preserve scenic and recreational values."[51] At the time, a lone gas well was operating within the protected area, and it continues to be a source of air pollution for the Uinta Basin.

In October 2015, following more than three years of talks between the inter-tribal coalition and the Utah Congressional delegation, Representatives Jason Chaffetz and Rob Bishop produced a compromise bill that would have desig-nated 1.2 million acres as a "national conservation area." The coalition balked at the proposal because the protected area was smaller than they had envisioned, and they feared the remainder of the Red Rock Wilderness would be open for accelerated of coal and oil shale development. Tribal leaders also objected because they would have a diminished role in managing the land under the proposed plan. Moreover, it would have been problematic for the Utah delegation to introduce their proposal as legislation. Following the controversy over the Grand Staircase-Escalante NM, and concerns about Obama's use of the Antiquities Act, state lawmakers passed a resolution declaring that the state's Congressional dele-gation should support no land preservation proposal that did not originate with the home County. Not only was the Chaffetz-Bishop bill unpopular with the majority-Navajo population of San Juan County, the County Commission was already on record as supporting the more expansive preservation plan.

When the coalition felt they were getting nowhere, they followed the example of the Southern Utah Wilderness Alliance and turned to the White House. With one year remaining in Obama's presidency, the coalition gave up on a legislative solution and presented the administration with a petition to have all 1.9 million acres protected, along with a plan to co-manage the proposed national monu-ment with the BLM. Over the following year, the Interior Department solicited comments and studied the proposal to confer monument status, a solution which would satisfy the coalition as well as the requirements of the public lands inven-tory. Conservative lawmakers took up the cause of ranchers and other rural resi-dents who opposed the plan, and Governor Gary Herbert visited the White House, hand-delivering a note urging Obama not to declare a new monument.[52] The Interior plan was to preserve 1.35 million acres, slightly larger than the conservation area proposed by the Utah congressional delegation. It was also a smaller area than the Inter-Tribal Coalition sought, but it formalized their role in the co-management of the monument. Obama issued his proclamation on December 28, 2016. Declaring Bears Ears "a milieu of the accessible and obser-vable together with the inaccessible and hidden," he announced the new national monument at the same time as Gold Butte NM in Nevada. Herbert called the proclamation "an arrogant act by a lame-duck president," and vowed to press Congress and the Trump Administration to undo it.

Downsizing Monuments

Soon after taking office, Trump ordered a review of the 27 largest national monuments created or expanded since Grand Staircase-Escalante in 1996.[53] The majority were located in Western states and designated by Clinton and Obama, although the list included the marine monuments created by George W. Bush.

Interior Secretary Ryan Zinke spent four months visiting the sites and attending public hearings. In an approach that resembled the Bush-Cheney model of collaborative conservation, participation was weighted in favor of groups with development interests in the land.

No president had ever rescinded a national monument designation by a predecessor. The history of the Antiquities Act has been one of presidents preserving land through unilateral action, and Congress reacting to those initiatives. In many cases, Congress upgrades national monuments to national park status. At other times, these presidential initiatives go unchallenged, and Congress validates them with its tacit approval. Some national monuments have been reduced in size through legislative action, but the pattern has been one of presidents acting unilaterally to preserve land, with Congress acting as a check on this power. There had been cases of presidents unilaterally changing the boundaries of national monuments, but because no there had been no legal challenges, the courts never ruled on the constitutionality of those actions.[54] The justification came from a paper by John Yoo and Todd Gaziano, which argued that presidents should not have their hands tied by actions of previous presidents. Despite the absence of precedent, they advised the Trump Administration that any grant of discretionary power includes the authority to reverse it.[55]

The outcome of the National Monument Review was hardly in doubt because Zinke had stated his intent to reduce the size of Bears Ears before the review began. This change was sought by Energy Fuels Resources USA, a Canadian-owned uranium company.[56] In August 2017, he recommended trimming Bears Ears to 16 percent of its original size. He also recommended reducing Grand Staircase-Escalante by nearly half, potentially unlocking significant coal reserves. This particular recommendation was unexpected because the controversy over this monument had subsided long ago: In 1998, the State of Utah and the federal government made a land swap, which at the time, satisfied Utah's governor and Congressional delegation. Two other land monuments, Nevada's Gold Butte and Oregon's Cascade-Siskiyou, also were designated for reductions in size.

It was later revealed that the Interior department discounted the benefits of the national monuments during the review. A set of unredacted documents accidentally released to journalists included communications from the department's Freedom of Information Act (FOIA) officers, indicting they sought to withhold any information that could contradict the administration's push to reverse or revise national monument designations. Department officials were told to emphasize the economic benefits of mining, logging, grazing, and energy production, while disregarding the benefits of tourism, recreation and archaeological discoveries on the sites.[57] Several weeks after the embarrassing document release, the Interior Department announced a change in procedures. In his last week before resigning, and amid a government shutdown, Zinke submitted a proposed rule to relax the timeline the agency follows to respond to FOIA requests.[58]

George W. Bush's preservation legacy was another casualty of the review. Significant reductions were recommended for the Pacific Remote Islands and Rose Atoll National Marine Monument, both of which were dedicated before Bush left office in 2009. Finally, the review designated a reduction for the lone Atlantic Ocean marine monument, Northeast Canyons and Seamounts. This area of undersea rock formations and cold-water corals near Cape Cod was relatively tiny, compared to the Pacific monuments. It was set aside during Obama's second term, over the objections of the Massachusetts Lobstermen's Association, which wanted it to remain open to fishing. From the Interior Department's accidental document release, it was revealed that the lead staff member for the review had deleted language noting that fishing vessels active near the Northeast Canyons and Seamounts "generated less than five percent or less" of their annual catch from the protected area.[59]

Rangeland Policy

Much like federal timber and mining policies, federal grazing policy was the product of a broader effort to encourage development of natural resources but resulted in unsustainable practices. According to Charles Davis, grazing policy is a classic example of distributive policy because it is determined largely by a narrow set of interest groups that enjoy the benefits whereas the burdens are diffused widely among the taxpayers.[60] This powerful subgovernment has managed to fend off legislative reform through the influence in Congress it maintains. In another regard, federal rangeland management resembles National Forest management because it loses money. A GAO report estimated that federal agencies spend $140 million annually to support grazing, whereas grazing fees generate $21 million.[61]

Federal grazing lands were largely depleted by the time the Taylor Act enshrined a set of rancher-friendly management practices in the 1930s. That law created grazing advisory boards, dominated by livestock producers, slashed grazing fees and explicitly declared grazing as the predominant use of federal range lands, which encouraged overgrazing for decades. Congress attempted to address this in 1976 with the FLPMA, which required the BLM to balance development with ecological considerations. The result was less of a balance than a tense stalemate between competing interests.

Prior to his election as president, Ronald Reagan proclaimed his support for ranchers who demanded that ownership of federal lands be transferred to the states, in what became known as the Sagebrush Rebellion. Despite these overtures, Regan avoided legislative changes and used administrative channels to reorient the Interior Department's mission. He cut the budget for the BLM and appointed loyalists to the Interior Department committed to maximizing productivity of the range.[62] Furthermore, Secretary James Watt moved aggressively to shuffle personnel, placing Reagan's supporters in positions of responsibility. This approach tilted range management policy in favor of ranchers for the duration of the Reagan years, but it was a tentative solution because a future president could potentially use the same methods to reorient policy toward different goals.

When Bill Clinton was elected president, he chose as his Secretary of the Interior Bruce Babbitt, a former Arizona governor and rancher who identified land reform as his top priority. Clinton and Babbitt supported a bill in the 103rd Congress by Representative Mike Synar that would have increased grazing fees and replaced the grazing advisory boards with more diverse panels that included environmental groups and scientists, in accordance with the multiple-use mission of the BLM. Due to stiff opposition from Western delegations, that and other grazing reform bills in the104th and 105th Congresses failed. The administration thus was forced to find creative solutions using its existing authority.

In 1995, Babbitt proposed a package of administrative reforms that established regional resource advisory councils to develop range management plans for ecosystem preservation and to establish grazing guidelines. Importantly, environmental groups were guaranteed representation on the councils. Furthermore, the rules clarified that the federal government retained title to physical improvements on the rangelands made by ranchers. Opponents in the 104th Congress tried to preempt the Interior Department, but their bills were doomed by gridlock. A House bill, which never made it to a floor vote, would have increased fees by a nominal amount but also lengthen the terms of grazing permits and bar the Interior Secretary from setting national grazing standards. A senate bill would have established 150 rangeland advisory boards whose membership included ranchers only, but the proposal lacked the 60 votes necessary to overcome a likely Democratic filibuster.

Whereas opponents were unable to muster the votes for preemptive legislation, the administration moved forward decisively and issued final rules. Babbitt moved ahead with other administrative reforms, introducing ecosystem management as the criteria for evaluating rangelands, replacing a vague four-point classification scheme (excellent, good, fair and poor) which the BLM had used for decades.[63] In a brief setback, the new restrictions placed upon ranchers were ruled illegal in U.S. District Court in Wyoming on grounds that they exceeded the BLM's authority.[64] On appeal, both the Tenth Circuit Court and U.S. Supreme Court rejected major parts of the lower court ruling, and upheld that the BLM had the right to limit the number of cattle under a grazing permit.

Ranchers hoped for a better deal when George W. Bush became president. In 2003, the administration delivered. New grazing rules were proposed to reduce BLM discretion, restricting its monitoring and enforcement capabilities. The rules also reversed a provision of Babbitt's reforms, asserting that ranchers could be granted ownership of physical improvements made to rangeland such as fences and wells. After they were finalized in 2006, the rancher-friendly rules were struck down because of inadequate consideration of their environmental impact and limited opportunities for public participation in the rulemaking. In addition, the BLM was faulted for ignoring information from the EPA, FWS and its own scientists. The ruling, which held that the rules were in violation of NEPA, the ESA and FLPMA, was upheld in upon appeal to the Ninth Circuit in 2010. A decade after the U.S. Supreme Court upheld the administrative reforms made by the Clinton Interior Department, another president issued rules to undo most of the changes administratively.

Rangeland policy has remained a political football. After the Bush-era rules were overturned, the Obama Administration's Interior Department went to work on new regulations to modernize the planning process. The rulemaking continued through Obama's second term, under the guidance of Secretary Sally Jewell. A former executive of REI, Jewell's appointment as interior secretary was a sign of the administration's commitment to multiple use as a guiding principle; signaling that recreational uses of public land were as important as energy and resource development. The "Planning 2.0 Rule" required the BLM to increase public involvement and incorporate the most current technology and data to determine whether (and where) mining, drilling, and grazing would occur. Some Western politicians did not see the new regulations as increasing efficiency and transparency, but as creating greater access points for federal agencies, scientists, and environmentalists, and diminishing the influence of state and local governments.[65]

Because the new planning rule took effect in December of 2016, it was vulnerable to legislative repeal. The Congressional Review Act had been invoked to overturn a regulation only one time in its 20-year history, but the election of Trump with a Republican Congress meant every new regulation in the Federal Register was in danger. A repeal resolution sponsored by Wyoming Representative Liz Cheney passed the House easily. It was approved by a 51–48 vote in the Senate and signed by President Trump, killing the planning rule. Six days later, Trump signed another CRA resolution repealing regulations on hunting bears and wolves in Alaskan wildlife preserves.[66]

Mining Law Reform

As with grazing policy, overhauling the Mining Act of 1872 was another priority for Clinton and the 103rd Congress. That law was intended to spark development in the West by encouraging mining of precious metals. The environmental impact is severe because gold, silver and copper mining generate acid and cyanide runoff, and the open-pit mines consume significant amounts of additional land to support the operations. As with the federal grazing policy, the FLPMA did little to curtail the environmental degradation and loss of revenue because the practice of leasing federal land at below-market prices continued.

There was a growing consensus for reform of the 120-year-old mining law, a goal shared by environmentalists as well as free-market economists. Like the grazing law, hardrock mining reform appeared to be an idea whose time had come. Environmentalists who pressed for reform found new allies as the federal deficit became an issue, since the low mining fees set by the antiquated law became harder to justify. The administration backed a bill to raise royalties for minerals extracted from federal lands. However, Congressional delegations from Western states were united in opposition, and the reform bill lost by single vote. In 1995, Babbitt held an auction in which he defiantly sold off federal land to foreign mining firms at pennies per acre.

Clinton turned from legislation to rulemaking after the 103rd Congress foiled his attempt to raise royalties. Abandoning any hope of reform legislation in the Republican-controlled 104th Congress, the Interior Department developed new rules requiring prior approval for hardrock mining activity. Under the rules, if mine operators failed to show that they could avoid significant disturbance to the land, they could be denied permits by the Secretary of the Interior for purely ecological reasons. There was a consensus within the administration that the rulemaking was justified because of the failure of Congress to deliver a remedy to the problem through legislation.[67] The FLPMA granted the Secretary of the Interior the authority to prevent what it deemed "unnecessary or undue degradation" of the environment. The Clinton Administration rule clarified the meaning of this phrase, specifying conditions under which a mining permit could be denied.[68]

The new hardrock mining rule took effect on Clinton's final day in office. Incoming Interior Secretary Gale Norton promptly initiated a new rulemaking process by proposing new regulations that deleted the clarifying language.[69] However, there remained a broad bipartisan consensus in favor of mining law reform, and many Congressional Republicans as well as Democrats supported the Clinton Administration's mining regulations. Ironically, the short-lived Democratic majority in the U.S. Senate was instrumental in undoing the Clinton rule after the defection of Republican-turned-Independent Senator Jim Jeffords in 2001. After the House of Representatives voted to prevent the Department of Interior from repealing the rule, the Senate stripped that provision from the Interior appropriations bill. The charge against the provision was led by none other than the new Majority Whip, Harry Reid, representing the mining state of Nevada.[70] Soon after that, the Bush Administration's Interior Department completed a new rulemaking that restored the earlier procedures.

Reid was senate majority leader eight years later, when the Obama Administration attempted a more modest reform. The EPA in 2016 proposed a hardrock mining rule, which required mine operators to prove they could afford to pay for the cleanup of their own operations before permits were granted. It probably seemed like a sensible reform, after the Zortman-Landusky gold and silver mine in Montana went bankrupt in 1998, leaving taxpayers on the hook for the mitigation of toxic pollution of nearby streams and groundwater.[71] In 2017, Trump EPA Administrator Scott Pruitt withdrew the proposed rule. Pruitt reasoned, "Additional financial assurance requirements are unnecessary and would impose an undue burden on this important sector of the American economy and rural America," and the reversal effectively transferred the cost of future cleanups back to the taxpayers.[72]

Energy Development

A top priority of the Bush-Cheney Energy Strategy was to open more public land to oil and gas production. To accomplish this, the administration aggressively sold leases on federal land and sought to open the Arctic National Wildlife Refuge (ANWR) to drilling. Since the latter goal could not be accomplished under the administration's own authority, Bush relied on Congress to pass legislation to lift the prohibition on drilling in ANWR. Congress came close in 2005, at a time when Bush benefited from the largest Republican majorities of his presidency, when both houses approved bills to authorize drilling. Ultimately, the provision authorizing drilling in ANWR was dropped from the final bill by a group of moderate Republicans. The failure to open ANWR notwithstanding, the Bush Administration was able to expand petroleum production on federal lands under its existing authority. During the Clinton years, there was a rough balance between conservation and energy production. The ratio of land leased for oil and gas to land preserved of roughly one-to-one. Conservation fell far behind under President Bush, with administration leasing 7.5 acres for oil and gas for every acre set aside for preservation and recreational uses.[73]

Developing oil shale was another goal of the Bush-Cheney Energy Strategy. There had been attempts to lease public land for oil shale during the Reagan and George H.W. Bush years, but Congress blocked those amid concerns about environmental impact and unproven technology. In accordance with the administration's energy strategy, Congress passed the Oil Shale, Tar Sands, and Other Strategic Unconventional Fuels Act when it was under Republican leadership. That 2005 law authorized the sale of permits for oil shale mining to meet increasing energy demands. However, there was not great enthusiasm for developing this resource, even by the petroleum industry, because there was still no cost-effective method to extract hydrocarbons from shale. Furthermore, it was estimated that two to five barrels of water was needed for every barrel of oil produced.

Those efforts became more aggressive at the end of Bush's presidency, when it appeared that he would be succeeded by Obama. On Election Day, 2008, the administration auctioned leases for hundreds of thousands of acres in Utah. Even though the plan included parcels of land adjacent to national parks, the National Park Service was not told about the plan ahead of time.[74] The Bush Administration opened 2 million acres for oil shale leasing in Wyoming, Utah and Colorado at the end of 2008.[75] The largest sale occurred during the Bush presidency's final week, when the BLM issued a single lease of 640 million acres for exploration, with a long-term option and a low royalty rate of 5 percent.

In a clear break with the previous administration's priorities, Obama's Interior Department cancelled the leases from Bush's midnight auction. Secretary Salazar established offices of renewable energy in four Western states to process permits for renewable energy projects on federal land, and he issued a directive to set aside 676,000 acres of BLM land for solar energy study zones.[76] The new administration was just as active in leasing for oil and gas development, but Obama resisted demands to open ANWR by focusing attention elsewhere in Alaska. Fulfilling a campaign promise, he directed the Interior Department to conduct a review of the National Petroleum Reserve. In 2011, Salazar released the NPR-A plan, which balanced development and conservation goals while preserving native land claims. At the time Obama took office, the natural gas industry was invigorated by new technologies that reduced the cost of extraction. Even though much of this development took place on private land, he continued his predecessor's practice of leasing public land for natural gas, which was a low-cost, cleaner-burning alternative to oil and coal. This energy agenda created tension with Obama's own environmental goals, and nothing illustrates this better than the controversy over the gas industry's practice of hydraulic fracturing. As I argue in Chapter 5, environmentalists accused the administration of looking the other way until the EPA was forced to confront the problem of hazardous pollution generated from this practice. Obama's "all of the above" strategy became a political liability at times: the Deepwater Horizon disaster in the Gulf of Mexico occurred shortly after he endorsed offshore oil production on the Outer-Continental Shelf and elsewhere his early advocacy for nuclear power appeared misguided after the Fukushima disaster in Japan.

The Trump Administration was arguably a greater champion of hydrocarbon development in Western lands than the Bush-Cheney Administration had been. This included leasing public lands for coal mining. In 2017, when Western public lands accounted for 40 percent of coal production, Trump issued an executive order reversing his predecessor's three-year moratorium on new coal leases. The Interior Department put so many acres up for oil and gas leasing, that there were often more parcels than bids. Consequently, many leases were given away for tiny sums of money. Parcels that did not receive initial bids were put up for non-competitive bids, which speculators purchased for as little as $1.50 per acre.[77] By comparison, the average bid during the last two years of the Obama Administration was $100. The administration's dedication to the petroleum industry was unbroken after Zinke's departure on January 2, 2019. His deputy, David Bernhardt, became acting secretary at the start of a five-week government shutdown. The office responsible for oil and gas leasing continued to operate, even as Interior Department personnel deemed "nonessential," including employees of the National Park Service and Fish and Wildlife Service, were furloughed. During the shutdown, a number of parks and wildlife refuges were subject to degradation through neglect and vandalism.

Conclusion

This case study of public land and natural resource management suggests it matters who the president is. It matters especially who the top administrators federal natural resource agencies are, because they wield significant influence over agency policies in the absence of clear directions from Congress. Legislative debate is deadlocked because of persistent pressure from mining and timber interests on one side and environmentalists on the other. During the era of Congressional dominance, major legislation was enacted to set aside wilderness areas and to require land management agencies to balance development with wildlife conservation and recreational uses, but the effect of these statutes was to push conflict into the courts, and ultimately, the administrative arena. The absence of mass protests and voting on the basis of conservation issues suggest that the public at large is disengaged from land management policy debates, giving Congress little reason to consider interests outside of the narrow issue network. Consequently, the legislative process has become an ineffective means to manage conflict. Interest groups increasingly take their demands directly to the bureaucracy, where the president wields the most influence.

When political conflict is displaced, agencies such as the Bureau of Land Management and the Forest Service make significant policy decisions through the rule-making process.[78] This makes control of the executive branch critical. The party that wins the White House has a chance to do this if the president uses the tools of administration effectively. The example of the Forest Service illustrates the tendency of policy to change direction when leadership changes. Promoting timber harvesting for the forest products industry is one of its mandates, but ambiguity over how to balance those competing mandates enhances administrative discretion. Political appointees and managers were thus empowered to introduce ecosystem-based management as a guiding principle for implementation. This approach has been called problematic because it prioritizes preservation of ecosystem integrity over other uses, whereas multiple use is more consistent with the politics of pluralism. Such a sweeping policy change would be more effective and legitimate if produced by the legislative process, not left to the discretion of managers.[79]

Instead of resolving lingering questions of whether to harvest timber, how much and where, policymakers rely upon the administrative arena to manage conflicts over environmental impact studies, scientific reviews, scoping for new rules, administrative appeals and road construction. Managers of the bureaucracy find themselves deciding issues that the legislative process is unable to resolve. As the following chapter will argue, emerging problems such as nonpoint-source water pollution are likely to remain unaddressed in the absence of direction from the top.

Notes

1 The lead ammunition ban was a rule from the final hours of the Obama presidency, but not yet in effect. See Valerie Volcovi, 2017, "New Interior Head Lifts Lead Ammunition Ban in Nod to Hunters," Reuters (March 2).

2 See: Daryl Fears and Dino Grandoni, 2018, "The Trump Administration Officially Clipped the Wings of the Migratory Bird Treaty Act," *Washington Post* (April 13); Dan Weikel, 2017, "Trump Administration Cancels Proposed Limits on Marine Mammals and Sea Turtles Trapped in Fishing Nets," *Los Angeles Times* (June 12); Zaz Hollander, 2017, "Trump and Congress Revoked the Predator-Control Ban on Alaska's Refuges," *Anchorage Daily News* (April 10).

3 David Bernhardt, 2018, "At Interior, We Are Ready to Bring the Endangered Species Act Up to Date," *Washington Post* (August 9).

4 Jim Tankersley, 2009, "Obama Overrides Bush on Endangered Species Act," *Los Angeles Times* (November): A1.

5 The editorial quoted Dirk Kempthorne, Secretary of the Interior in the Bush Administration, as boasting that the rule would prevent the ESA from being used as a "back door" to block future construction projects. See *New York Times*, 2008, "An Endangered Act" (August 13).

6 Union of Concerned Scientists, "Scientific Integrity in Policymaking" (February 2004), www.ucsusa.org.

7 Ibid.

8 In the markup, powerful Alaska Senator Ted Stevens inserted vague compromise language providing for greater participation by the advisory councils. This language presented an opportunity for a clarification in the form of a rule change. See Erika Lovely and Ryan Grim, "Bush Fights to Loosen Fishery Rules," *Politico* (December 9, 2008), www.politico.com.

9 Joe Roman, 2011, *Listed*, Cambridge, MA: Harvard University Press.

10 Lisa, Friedman, Kendra Pierre-Louis and Livia Albeck-Ripka, 2018, "Law That Saved the Bald Eagle Could be Vastly Reworked," *New York Times* (July 19).

11 Walter A. Rosenbaum, 2012, *Environmental Politics and Policy*, 8th edition, Washington, DC: Congressional Quarterly Press.

12 Richard Tobin, 1990, *The Expendable Future: U.S. Politics and the Protection of Biological Diversity*, Durham, NC: Duke University Press.

13 The USFS had a reputation as a model bureaucracy for these reasons. See Jeanne N. Clarke, and Daniel McCool, 1985, *Staking Out the Terrain: Power Differentials Among Natural Resource Management Agencies*, Albany: SUNY Press.

14 George Hoberg, 1997, "From Localism to Legalism: The Transformation of Federal Forest Policy" in *Western Public Lands and Environmental Politics*, edited by Charles Davis, Boulder, CO: Westview: 47–73.

15 Ibid.

16 Steven L. Yaffee, 1994, *The Wisdom of the Spotted Owl*, Washington, DC: Island Press.

17 Judith. A. Layzer, 2012, *The Environmental Case: Translating Values to Public Policy*, 3rd edition, Washington, DC: Congressional Quarterly: 455

18 Sarah B. Praelle, 2006, *Branching Out, Digging In: Environmental Advocacy and Agenda-Setting*, Washington, DC: Georgetown University Press.

19 Barbara Sinclair, 2007, *Unorthodox Lawmaking*, 3rd edition, Washington, DC: Congressional Quarterly Press.

20 Quoted in *Seattle Post-Intelligencer*, 1995, "White House Blunders on Environmental Bills," (December 4).

21 Tom Turner, 2009, *Roadless Rules: The Struggle for the Last Wild Forests*, Washington, DC: Island Press.

22 Two earlier reviews of roadless areas, RARE (1973) and RARE II (1979) both resulted in recommendations to suspend road construction. RARE II was considered more thorough and with a stronger basis in conservation science and appeared to have a stronger chance of successful implementation, but it was blocked by Congress. See Michael P. Dombeck, Christopher A. Wood and Jack E. Williams, 2003, *From Conquest to Conservation*, Washington, DC: Island Press.

23 Four alternatives were considered. The first was to maintain the status quo, which assumed the forest service could find a way to maintain its roadbuilding and repair obligations while meeting timber quotas. The other three alternatives banned road construction and replacement in inventoried areas, but they varied in their restrictions on timber harvesting. Those alternatives ranged from a prohibition on all timber harvesting (except for the purpose of habitat protection) to no restriction at all. Finally, in between these extremes, there was one more alternative that allowed timber harvesting

in limited circumstances. This compromise is the alternative that was ultimately recommended. See Tom Turner, 2009, *Roadless Rules: The Struggle for the Last Wild Forests*, Washington, DC: Island Press.

24 Christopher McGrory Klyza and David Sousa, 2008, *American Environmental Policy, 1990–2006*, Cambridge, MA: MIT Press.

25 Bill Clinton, "Remarks on Action to Preserve America's Forests," January 5, 2001, *Public Papers of the President: William J. Clinton: 2001*, Washington, DC: 18.

26 Andrew H. Card, Jr., 2001. "Memorandum for the Heads and Acting Heads of Executive Departments and Agencies; Subject: Regulatory Review Plan." 66 F.R. 7702 (January 24).

27 Jacqueline Vaughn and Hanna Cortier, 2004, "Using Parallel Strategies to Promote Change: Forest Policymaking Under George W. Bush," *Review of Policy Research* 21: 767–782.

28 The administration essentially took the plaintiffs' side, conceding that there had not been "sufficient and meaningful opportunity" for participation during the scoping phase. *Kootenai Tribe of Idaho v. Veneman*, 313 F.3d 1094 (9th Cir. 2002).

29 In 2009, the Bush rule was remanded when environmentalists appealed the decision in the Idaho case to the Ninth Circuit Court. The ruling reinstated the Clinton rule, but only for the Ninth District's jurisdiction, as the ruling was in direct conflict with the ruling in the Wyoming case. See Tom Turner, 2009, *Roadless Rules: The Struggle for the Last Wild Forests*, Washington, DC: Island Press.

30 Martin Nie, 2008, *The Governance of Western Lands: Mapping Its Present and Future*, Lawrence, KS: University Press of Kansas.

31 Vilsack's directive required all new road construction projects to be cleared through his office, in order to "provide consistency and clarity that will help protect our national forests until a long-term roadless policy reflecting President Obama's commitment is developed." See U.S. Department of Agriculture, 2009, News Release, "Agriculture Secretary Vilsack Announces Interim Directive Governing Roadless Areas in National Forests," Document 0185.09 (May 28).

32 James Skillen, 2015, *Federal Ecosystem Management*, Lawrence, KS: University Press of Kansas.

33 Martin Nie argues the ban was grounded in science, but nonetheless a political decision. The timing of the decision was strategic. By listing the final rule on November 22, 2000, the administration ensured it would take effect 60 days before the end of Clinton's term. With the outcome of the presidential election still in dispute, it was unclear whether he would be succeeded by Bush or Gore. The timing precluded Clinton's successor from rescinding the ban without initiating a new rulemaking process. See Martin Nie, 2008, *The Governance of Western Lands: Mapping Its Present and Future*, Lawrence, KS: University Press of Kansas.

34 National Park Service, 2000, "Winter Use Plan: Record of Decision," F.R. 65 (November 22): 80,908.

35 Diego Stein, "Snowmobile Ban Set for '03 at 2 Parks," *Denver Post* (November 23, 2000): A1.

36 The proposed replacement rule explicitly declared that snowmobile recreation was an acceptable use of the parks, F.R. 68 (December 11, 2003): 69,268.

37 The administration ultimately reached a settlement, successfully arguing that the technology was the issue. Since most personal snowmobiles featured noisy high-emission, two-stroke engines, only private snowmobiles would be prohibited. Under the revised Winter Use Plan in 2007, licensed tour companies would be permitted to operate snowmobiles in the parks. See Martin Nie, 2008, *The Governance of Western Lands: Mapping Its Present and Future*, Lawrence, KS: University Press of Kansas.

38 Ken Salazar, Memorandum, "Wilderness Policy" (June 1, 2011), www.doi.gov/news/pressreleases/upload/Salazar-Wilderness-Memo-Final.pdf.

39 See, for example, Robert G. Hillman, "At Grand Canyon, Clinton Touts Environmental Record," *Dallas Morning News* (September 19, 1996): A10.
40 Peter Hoagland, 1995, Letter to George Stephanopolous, William J. Clinton Presidential Library: Presidential Records, George Stephanopolous OA/ID Number 10168.
41 The monument designation was viewed with more cynicism in the region. Former Montana House of Representatives Speaker Daniel Kemmis, a democrat, wrote, "... from the national Democratic Party's perspective, the loss of this lone Democrat in the Utah delegation was a reasonable price to pay for the votes and support of environmental groups that saw the creation of the monument as a major victory. Undeniably, these actions have lost Democrats support among Westerners. But then, these are the people who essentially do not exist in the tactical calculations of the national Democratic Party. See Daniel Kemmis, 2001, *This Sovereign Land*, Washington, DC: Island Press.
42 Tom Turner, 2009, *Roadless Rules: The Struggle for the Last Wild Forests*, Washington, DC: Island Press.
43 Monument-creation intensified at the very end of his final year. During January 2001, his last three weeks as president, Clinton created or expanded nine national monuments. See Sanjay Ranchod, 2001, "The Clinton National Monuments: Protecting Ecosystems with the Antiquities Act," *Harvard Law Review* 25: 535–589.
44 Bush used proclamations to designate or add to four Marine National Monuments in the Pacific. These were the Papahanaumokuakea in the Northwest Hawaiian Islands, the Marianas Trench, Pacific Remote Islands, and Rose Atoll Marine National Monuments.
45 Kirk Johnson, "In the West, 'Monument' Is a Fighting Word," *New York Times* (February 20, 2010): A8.
46 Bureau of Land Management, "Treasured Landscapes," Discussion Paper (no date).
47 Senator Jim DeMint, "White House Land Grab; Proposal to Seize Land Would Favor Animals Over Americans". See Jim DeMint, "White House Land Grab; Proposal to Seize Land Would Favor Animals Over Americans," *Washington Times* (March 3, 2010): B1..
48 In early 2013, he declared four more monuments of historic and ecological significance, ranging in size from small national historic monuments honoring settlers in Delaware, Buffalo Soldiers in Ohio and Harriet Tubman and the Underground Railroad in Maryland, to the 240,000-acre Rio Grande Gateway in New Mexico and the 1,000-acre San Juan Islands National Monument in Washington State. See John Broder, "Obama to Name New National Monuments," *New York Times* (March 22, 2013).
49 Steve Tetreault and Henry Brean, 2015, "A Done Deal, Obama to Create Basin and Range National Monument," *Las Vegas Review-Journal* (July 9).
50 Stephen Trimble, 2016, "One Well: Drilling the Bears Ears" in *Fracture: Essays, Poems and Stories on Fracking in America*, edited by Taylor Brorby and Stefanie Brook Trout, North Liberty, IA: Ice Cube Press, 141–149.
51 Ibid.
52 Jack Healy, 2016, "Remote Utah Enclave Becomes New Battleground over Reach of U.S. Control," *New York Times* (March 12): A1.
53 Donald J. Trump, "Executive Order 13792—Review of Designations Under the Antiquities Act." Online by Gerhard Peters and John T. Woolley, The American Presidency Project, www.presidency.ucsb.edu/node/326685.
54 For example, President Woodrow Wilson drastically shrunk Washington's Olympia National Monument before it was given national park status by Congress, and President Kennedy both removed acreage from and added to Bandalier National Monument in New Mexico.
55 The paper was later published as a report by the American Enterprise Institute. See John Yoo and Todd Gaziano, "Presidential Authority to Revoke or Reduce National Monument Designations," American Enterprise Institute, March 2017.

56 Juliet Eilperin, 2017, "Uranium Firm Urged Trump Officials to Shrink Bears Ears National Monument," *Washington Post* (December 8).

57 Juliet Eilperin, 2018, "Trump Administration Officials Dismissed Benefits of National Monuments," *Washington Post* (July 23).

58 43 CFR Part 2 [Docket No. DOI-2018–0017] RIN 1093-AA26 Freedom of Information Act Regulations (July 28, 2018):67175–81.

59 Quoted in Juliet Eilperin, 2018, "Trump Administration Officials Dismissed Benefits of National Monuments," *Washington Post* (July 23).

60 Charles Davis, 1997, "Politics and Rangeland Policy" in *Western Public Lands and Environmental Politics*, edited by Charles Davis. Boulder, CO: Westview. 74–94.

61 U.S. Government Accountability Office, 2005, "EPA Should Devote More Attention to Environmental Justice When Developing Clean Air Rules," GAO-05–289 (July), www.gao.gov/new.items/d05289.pdf.

62 Robert F. Durant, 1992, *The Administrative Presidency Revisited: Public Lands, the BLM, and the Reagan Revolution*, Albany: SUNY Press.

63 Bruce Babbitt, 2000, Remarks to BLM Employees in Phoenix, Arizona, April 24.

64 The opinion was issued by Judge Clarence Brimmer, the same judge who later struck down the roadless rule.

65 Bruce Finley, 2017, "House Votes to Kill Planning 2.0 Rule," *Denver Post* (February 7).

66 The rules prohibited the hunting of wolves during denning season and mother bears with cubs and banned the use of traps while hunting those predator species.

67 Interior Department Solicitor John Leshy argued that administrative policymaking does not usurp Congressional authority when the legislative branch refuses to take action. See John Leshy, 2001, "The Babbitt Legacy at the Department of the Interior: A Preliminary View," *Environmental Law* 31(2): 199–228.

68 The interpretation of the phrase became any "substantial irreparable harm to scientific, cultural or environmental resource values of public lands that cannot effectively be mitigated." 43 CFR Part 2090 "Mining Claims Under the General Mining Laws" (November 21, 2000): 69999.

69 The Bush Interior Department was sympathetic to the mining industry's argument that the "substantial irreparable harm" standard was itself too vague and overly broad to serve as the basis for denying a permit. See Douglas Jehl, "Gold Miners Eager for Bush to Roll Back Clinton Rules," *New York Times* (August 16, 2001): A1.

70 Dan Morgan, "New Mining Regulations Overturn Late Clinton Rules," *Washington Post* (October 26, 2001): A33.

71 Karl Puckett, 2018, "Fort Belknap Backs State in Bad Actor Case over Zortman-Landsky Pollution," *Great Falls Tribune* (September 13).

72 Ibid. Senator Tom Udall of New Mexico responded to the withdrawal by proposing legislation to clarify liability for pollution from hardrock mining, but the 115th Congress never considered his bill.

73 Bruce Babbitt, 2013, "On Equal Ground: Righting the Balance Between Energy Development and the Conservation of Public Lands," remarks at the National Press Club, Washington DC (February 5).

74 On December 19, 131 leases were auctioned. One of Salazar's first acts as Interior Secretary was to cancel 77 of the leases for sites adjacent to national parks and monuments in Utah. See Nicholas Riccardi and Jim Tankersley, 2009, "Salazar Cancels Bush-Era Energy Leases in Utah" (February 5): A10.

75 The BLM proposed a framework for oil shale development in July, and after four months on the fast-track, the final rule was issued on November 17. The timing ensured it would be in effect more than 60 days before Inauguration Day, safeguarding the rule from reversal from the incoming Obama Administration. See Jim Tankersley, 2008, "Bush Angers Environmentalists with Last-Minute Rule Changes," *Los Angeles Times* (March 4): A14.

76 Charles Davis, Sandra Davis, and Cassandra Koerner, 2016, "Renewable Energy Policymaking in the American West" in *Environmental Politics and Policy in the West*, 3rd edition, edited by Zachary A. Smith and John Freemuth. Boulder, CO: University of Colorado Press. 101–125.
77 Eric Lipton and Hiroko Tabuchi, 2018, "For Lease: Land to Drill on, Dirt Cheap," *New York Times* (November 28).
78 Klyza, Christopher McGrory and David Sousa, 2008, *American Environmental Policy, 1990–2006*, Cambridge, MA: MIT Press.
79 Sedjo, Roger, 1995, "Ecosystem Management: An Uncharted Path for Public Forests," *Resources* 121 (Fall).

Bibliography

Babbitt, Bruce. 2000. Remarks to BLM Employees in Phoenix, Arizona. April 24.
Babbitt, Bruce. 2013. "On Equal Ground: Righting the Balance Between Energy Development and the Conservation of Public Lands." Remarks at the National Press Club, Washington, DC (February 5).
Bernhardt, David. 2018. "At Interior, We Are Ready to Bring the Endangered Species Act Up to Date." *Washington Post* (August 9).
Broder, John. "Obama to Name New National Monuments." *New York Times* (March 22, 2013).
Bureau of Land Management. "Treasured Landscapes." Discussion Paper (no date).
Card, Andrew H., Jr. 2001. "Memorandum for the Heads and Acting Heads of Executive Departments and Agencies; Subject: Regulatory Review Plan." 66 F.R. 7702 (January 24).
Clarke, Jeanne N. and Daniel McCool. 1985. *Staking Out the Terrain: Power Differentials Among Natural Resource Management Agencies*. Albany: SUNY Press.
Clinton, William J. "Remarks on Action to Preserve America's Forests." January 5, 2001. *Public Papers of the President: William J. Clinton: 2001*. Washington, DC: 18.
Davis, Charles. 1997. "Politics and Rangeland Policy." In *Western Public Lands and Environmental Politics*, edited by Charles Davis. Boulder, CO: Westview. 74–94.
Davis, Charles, Sandra Davis, and Cassandra Koerner. 2016. "Renewable Energy Policymaking in the American West." In *Environmental Politics and Policy in the West*, 3rd edition. Edited by Zachary A. Smith and John Freemuth. Boulder, CO: University of Colorado Press. 101–125.
DeMint, Jim. 2010. "White House Land Grab; Proposal to Sieze Land Would Favor Animals Over Americans." *Washington Times* (March 3): B1.
Dombeck, Michael P., Christopher A. Wood and Jack E. Williams. 2003. *From Conquest to Conservation*. Washington, DC: Island Press.
Durant, Robert F. 1992. *The Administrative Presidency Revisited: Public Lands, the BLM, and the Reagan Revolution*. Albany: SUNY Press.
Eilperin, Juliet. 2017. "Uranium Form Urged Trump Officials to Shrink Bears Ears National Monument." *Washington Post* (December 8).
Eilperin, Juliet. 2018. "Trump Administration Officials Dismissed Benefits of National Monuments." *Washington Post* (July 23).
Fears, Daryl and Dino Grandoni. 2018. "The Trump Administration Officially Clipped the Wings of the Migratory Bird Treaty Act." *Washington Post* (April 13)
Finley, Bruce. 2017. "House Votes to Kill Planning 2.0 Rule." *Denver Post* (February 7).
Friedman, Lisa, Kendra Pierre-Louis, and Livia Albeck-Ripka. 2018. "Law That Saved the Bald Eagle Could be Vastly Reworked." *New York Times* (July 19).

Healy, Jack. 2016. "Remote Utah Enclave Becomes New Battleground over Reach of U. S. Control." *New York Times* (March 12): A1.

Hillman, Robert G. "At Grand Canyon, Clinton Touts Environmental Record." *Dallas Morning News* (September 19, 1996): A10.

Hoagland, Peter. 1995. Letter to George Stephanopolous, William J. Clinton Presidential Library: Presidential Records George Stephanopolous OA/ID Number 10168.

Hoberg, George. 1997. "From Localism to Legalism: The Transformation of Federal Forest Policy" in *Western Public Lands and Environmental Politics*, edited by Charles Davis. Boulder, CO: Westview: 47–73.

Hollander, Zaz. 2017. "Trump and Congress Revoked the Predator-Control Ban on Alaska's Refuges." *Anchorage Daily News* (April 10).

Jehl, Douglas. "Gold Miners Eager for Bush to Roll Back Clinton Rules." *New York Times* (August 16, 2001): A1.

Johnson, Kirk. "In the West, 'Monument' Is a Fighting Word." *New York Times* (February 20, 2010): A8.

Kemmis, Daniel. 2001. *This Sovereign Land*. Washington, DC: Island Press.

Klyza, Christopher McGrory and David Sousa. 2008. *American Environmental Policy, 1990–2006*. Cambridge, MA: MIT Press.

Kootenai Tribe of Idaho v. Veneman. 313 F.3d 1094 (9th Cir., 2002).

Layzer, Judith A. 2012. *The Environmental Case: Translating Values to Public Policy*, 3rd edition. Washington, DC: Congressional Quarterly: 455.

Leshy, John. 2001. "The Babbitt Legacy at the Department of the Interior: A Preliminary View." *Environmental Law* 31(2): 199–228).

Lipton, Eric and Hiroko Tabuchi. 2018. "For Lease: Land to Drill on, Dirt Cheap." *New York Times* (November 28).

Lovely, Erika and Ryan Grim. "Bush Fights to Loosen Fishery Rules." *Politico*. (December 9, 2008), www.politico.com.

Morgan, Dan. "New Mining Regulations Overturn Late Clinton Rules." *Washington Post* (October 26, 2001): A33.

National Park Service. 2000. "Winter Use Plan: Record of Decision." F.R. 65 (November 22): 80,908.

New York Times. 2008. "An Endangered Act" (August 13).

Nie, Martin. 2008. *The Governance of Western Lands: Mapping Its Present and Future. Lawrence*. Lawrence, KS: University Press of Kansas.

Praelle, Sarah B. 2006. *Branching Out, Digging In: Environmental Advocacy and Agenda-Setting*. Washington, DC: Georgetown University Press.

Puckett, Karl. 2018. "Fort Belknap Backs State in Bad Actor Case over Zortman-Landsky Pollution." *Great Falls Tribune* (September 13).

Ranchod, Sanjay. 2001. "The Clinton National Monuments: Protecting Ecosystems with the Antiquities Act." *Harvard Law Review* 25: 535–589.

Riccardi, Nicholas and Jim Tankersley. 2009. "Salazar Cancels Bush-Era Energy Leases in Utah." (February 5): A10.

Rosenbaum, Walter A. 2012. *Environmental Politics and Policy*, 8th edition. Washington, DC: Congressional Quarterly Press.

Roman, Joe. 2011. *Listed*. Cambridge, MA: Harvard University Press.

Salazar, Ken. Memorandum, "Wilderness Policy" (June 1, 2011)www.doi.gov/news/p ressreleases/upload/Salazar-Wilderness-Memo-Final.pdf.

Sedjo, Roger. 1995. "Ecosystem Management: An Uncharted Path for Public Forests." *Resources* 121 (Fall).

Sinclair, Barbara. 2007. *Unorthodox Lawmaking*, 3rd edition. Washington, DC: Congressional Quarterly Press.

Skillen, James. 2015. *Federal Ecosystem Management*. Lawrence, KS: University Press of Kansas.

Stein, Diego. "Snowmobile Ban Set for '03 at 2 Parks." *Denver Post* (November 23, 2000): A1.

Tankersley, Jim. 2008. "Bush Angers Environmentalists with Last-Minute Rule Changes." *Los Angeles Times* (March 4): A14.

Tankersley, Jim. 2009. "Obama Overrides Bush on Endangered Species Act." *Los Angeles Times* (November): A1.

Tetreault, Steve and Henry Brean. 2015. "A Done Deal, Obama to Create Basin and Range National Monument." *Las Vegas Review-Journal* (July 9).

Tobin, Richard. 1990. *The Expendable Future: U.S. Politics and the Protection of Biological Diversity*. Durham, NC: Duke University Press.

Trimble, Stephen. 2016. "One Well: Drilling the Bears Ears" in *Fracture: Essays, Poems and Stories on Fracking in America*, edited by Taylor Brorby and Stefanie Brook Trout. North Liberty, IA: Ice Cube Press. 141–149.

Trump, Donald J. "Executive Order 13792—Review of Designations Under the Antiquities Act." Online by Gerhard Peters and John T. Woolley, The American Presidency Project. www.presidency.ucsb.edu/node/326685.

Turner, Tom. 2009. *Roadless Rules: The Struggle for the Last Wild Forests*. Washington, DC: Island Press.

Union of Concerned Scientists. "Scientific Integrity in Policymaking." (February 2004). www.ucsusa.org.

U.S. Department of Agriculture. 2009. News Release, "Agriculture Secretary Vilsack Announces Interim Directive Governing Roadless Areas in National Forests." Document 0185.09 (May 28).

U.S. Government Accountability Office. 2005. "EPA Should Devote More Attention to Environmental Justice When Developing Clean Air Rules." GAO-05–289 (July).www.gao.gov/new.items/d05289.pdf.

Vaughn, Jacqueline and Hanna Cortier. 2004. "Using Parallel Strategies to Promote Change: Forest Policymaking Under George W. Bush." *Review of Policy Research* 21: 767–782.

Volcovi, Valerie. 2017. "New Interior Head Lifts Lead Ammunition Ban in Nod to Hunters." Reuters (March 2).

Weikel, Dan. 2017. "Trump Administration Cancels Proposed Limits on Marine Mammals and Sea Turtles Trapped in Fishing Nets." *Los Angeles Times* (June 12).

Yaffee, Steven L. 1994. *The Wisdom of the Spotted Owl*. Washington, DC: Island Press.

Yoo, John and Todd Gaziano. "Presidential Authority to Revoke or Reduce National Monument Designations." American Enterprise Institute (March 2017).

4

WATER QUALITY

The discovery of lead contamination in Flint, Michigan's tap water in 2015 should
have been a wakeup call. Nearly 8,000 children under six years of age were
exposed to unsafe levels of lead because state-appointed city managers began
drawing on the Flint River for drinking water in a cost-cutting move. The switch
was made without conducting tests required by the EPA's Lead and Copper Rule,
and it was 18 months before residents and volunteer researchers discovered the
water from the new source had corroded the city's aging lead pipes, exposing
residents to levels between 5,000 and 13,000 ppb (The federal standard for lead is
15 ppb).[1] A pediatrician who discovered the high lead concentrations wrote,
"Drinking Flint water after the switch was like drinking through a lead straw. You
never knew when a piece of severely corroding lead scale would break off and fall
into the drinking water and into the bodies of our children."[2] In short time, media
reports revealed that Flint was not unique, as communities around the country
found hazardous chemicals in their tap water. A similar crisis unfolded in Sebring,
Ohio, where officials failed to add chemicals to the town's drinking water to prevent the corrosion of lead pipes. Six months passed before warnings for children
and pregnant women not to drink the contaminated water.[3] Even though a few
politicians tried to make Flint into a symbol of a nationwide problem, there was
little awareness that an estimated 77 million Americans lived in communities with
one or more violation of federal standards.[4]

Water quality had been a top-tier issue four decades earlier, when Congress
nationalized the fight against water pollution, passing the Clean Water Act
(CWA) of 1972 and Safe Drinking Water Act (SDWA) of 1974. Reflecting on
that era, the EPA's first leader rendered a sobering assessment of the environmental and political challenges of the 21st century. Not only had the nature of
the problem changed, but there is also a drastically different political climate.

"When we started regulating water pollution in the 1970s, there was a huge public outcry because you could see raw sewage flowing into rivers," William Ruckelshaus recalled. "Today, the violations are much more subtle—pesticides and chemicals you can't see or smell that are even more dangerous…and so a lot of the public pressure on agencies has ebbed away."[5] The Clean Water Act successfully addressed the most publicly visible forms of water pollution, including untreated sewage and industrial wastewater. These first-generation problems were the most salient to the mass public, as well as the simplest to control.

The nation's rivers and streams appeared to be cleaner after five decades, and water quality became a lower priority for the public. By some indicators, water quality was declining due to pollution from agricultural and urban runoff. A comprehensive survey of the nation's rivers and streams completed in 2009 found only 34 percent of stream miles with "good" levels of phosphorus, down from 53 percent in 2004.[6] Without public pressure, Congress had little motivation to address the other water quality problems festering beneath the surface. This chapter examines the role of the executive bureaucracy in the promotion and diffusion of collaborative approaches to water pollution control and the retreat of Congressional leadership. Within government, advocates for policy solutions to the new problems were more often specialists within the bureaucracy, and, remarkably, presidents. The case study of water quality policy is particularly relevant to the present study's exploration of the environmental presidency, because it demonstrates the limits of first-generation policies as well as the challenges of some of the second-generation policies.

A role reversal between executive and legislative branches is particularly apparent in the case of water policy. For the two decades following the first Earth Day on April 22, 1970, Congress took the lead in developing policies to clean up the nation's rivers and streams and safeguard the supply of drinking water to better protect public health. Nixon was cautious about supporting water quality legislation, in spite of his enthusiasm for the Clean Air Act and the EPA. The Clean Water Act became law only after Congress overrode Nixon's veto in 1972. During the 1980s, the CWA was reauthorized and strengthened, despite twice being vetoed by President Reagan. Because water quality is most apparent to the public at the local level, members of Congress were sensitive to conditions in their districts and responsive to their constituents' concerns. When they believed the Reagan Administration acted too slowly to implement its programs, they used their oversight power to pressure the EPA and amended statutes to include detailed instructions and strict timetables to limit the agency's discretion.

The legislative branch's familiar role as leader on pollution control policy faded in the 1990s. Congress lost interest in major environmental legislation, as lawmakers felt pressure from economic interests to roll back existing regulations. This retrenchment began even before the arrival of the 104th Congress. Prior to the 1994 election and the Republican takeover of Congress, bills to strengthen the Clean Water and Safe Drinking Water Acts were shelved.[7] Congress launched a rearguard assault on the environmental policy regime after 1994 when there was a reversal of roles between the executive and legislative branches. The Clinton Administration championed new approaches to water policy, particularly in negotiating rulemaking and unifying rulemaking, while fending off legislative efforts to undermine the regulatory state. President Clinton used a variety of administrative tools to leave a mark on natural resource policy, including agency rulemaking, executive orders, veto threats, and signing statements, as detailed in the previous chapter. Executive rulemaking became a primary pathway to make policy, although most pollution rules were issued under statutory mandate or court order. Moreover, water quality was another area in which Clinton used administrative tools to advance his environmental agenda.

The gridlock of the Clinton years was followed by eight years of continued environmental policy retrenchment. No major environmental legislation was enacted during the presidency of George W. Bush. This, too, was a period of executive leadership. All significant environmental initiatives originated with the administration, including the "healthy forests" policy described in the previous chapter, as well as protective policies such as the EPA's new standards for diesel emissions. Thus, both leaders used executive power to affect environmental regulation in different ways. The consequences for water pollution will be explored below.

The New Politics of Water Pollution

The critique of first-generation environmental policies was familiar by the 1990s. Early policies that relied upon end-of-pipe technologies were less effective against second-generation problems. There were diminishing returns from further restrictions on industrial sources, and the threat from nutrients and microbial contamination was growing. The one-time symbol of the nation's ailing waterways, Ohio's Cuyahoga River, showed visible improvement after years of limiting industrial discharge. Despite this progress, the river suffered from persistent high levels of bacteria, caused by urban and agricultural runoff, as well as discharges of raw sewage.[8] Continued progress required that federal controls be extended over new areas in the public spotlight, such as stormwater and agricultural runoff.

The years that followed the environmental decade of the 1970s were marked with growing public dissatisfaction with the federal government's dominant role. Even Nixon spoke of a need for a more cooperative federalism, and over the next 20 years, there was a growing chorus of criticism that state and local authorities should assume greater responsibility.[9] However, there was little evidence of an improved relationship between the federal government and states on water policy. The framework of first-generation command-and-control policies relied on the states to implement their own permit systems, once standards were set by the EPA. States were allowed to seek primacy for their water pollution programs, but little progress was made toward devolving authority to the states before the 1990s. Meanwhile, the EPA's share of funding for state pollution programs declined between 1986 and 2000, even as the average annual cost of state environmental programs increased by nearly two-thirds.[10] The Clean Water Act authorized grants for sewer construction, which went a long way toward advancing the statute's goals. The federal share for those projects under the 1972 law was 75 percent, later reduced to 55 percent in 1981. The 1987 CWA Amendments replaced them with block grants for a state revolving loan fund.

The legacy of the first two decades of the CWA represents a sharp reduction in pollution from point sources, which include municipal sewer and industrial wastewater discharges, but not agricultural irrigation. The law's foundation is a system of technologically defined effluent limits on discharges from specific polluters. The law was passed at a time when public concern over the environmental crisis was at its highest, and Congress was willing to dismantle a water quality law based on ambient pollutant levels and state-level management that it passed just seven years earlier. Advocates including Wisconsin Senator Gaylord Nelson believed that only a federal program of uniform standards could improve the condition of interstate waterways such as the Great Lakes. Even those members of Congress who were skeptical about the enhanced federal role were reluctant to oppose the law.[11] The national mood in the early 1970s was predisposed to increased social regulation, so defying the environmental movement in that climate could prove to be politically risky. In an ominous sign, seven of the 12 incumbent members of Congress, labeled the "dirty dozen" by Earth Day organizer Dennis Hayes, were defeated in the previous midterm election.

The changing political climate was reflected in public opinion. According to the Gallup Poll, water quality remained at the top of the list of environmental problems, as perceived by the public. However, there was a long-term decline in the intensity of public concern after 1990. As Table 4.1 shows, the percentage of the public indicating they were concerned "a great deal" about pollution of rivers, lakes and reservoirs, declined from a peak of 72 percent in 1989 to 46 percent in 2011. Moreover, Table 4.1 illustrates that public concern about pollution of drinking water, which also peaked at 72 percent in 2000, declined to 50 percent in 2010 and 51 percent in 2011.

Despite the early success of the CWA, the rate of improvement slowed. This was followed by a period of regress in several watersheds, as fiscally stressed states devoted fewer resources to enforcement. By the time the CWA was up for renewal in the 103rd Congress, the bill came under attack, and was finally shelved in 1994. Following the Republican victory in the 1994 election, one of the early actions of the new Congress was an attempt to rewrite the Clean Water Act. The new battle lines were drawn quickly by the spring of 1995, when Clinton observed the 25th anniversary of the first Earth Day. In a forceful speech, Clinton used the event to criticize legislative proposals on water pollution, cost-benefit analysis and regulatory takings, which he promised to veto:

TABLE 4.1 Percentage of Respondents "Very Concerned" About Water Quality, 1991–2011

Pollution of Drinking Water	67	64	51
Pollution of Rivers, Lakes and Streams	67	58	46
Contamination of Soil and Water by Toxic Waste	62	58	48
Air Pollution	59	48	36
Loss of Tropical Rain Forests	42	44	34
Global Warming	35	33	25
Extinction of Plant and Animal Species	–	43	34
Urban Sprawl and Loss of Open Space	–	42	27

> On this Earth Day, let me pledge we will not allow lobbyists to rewrite our environmental laws in ways that benefit polluters and hurt our families, our children, and our future. Reform? Yes. Modernize? You bet. But roll back health and safety? No. Let DDT into our food again? Not on your life. Create more tainted water or toxic waste... Never.[12]

Clinton delivered the Earth Day address just three days after a memorable news conference, in which the president was laden with the burden of explaining to reporters why he believed the presidency was still relevant at all.[13] Clinton marked the Earth Day anniversary by presenting Gaylord Nelson with the Presidential Medal of Freedom and then cast himself as the defender of the retired senator's legacy. Moreover, the 104th Congress succeeded in relaxing restrictions on discharge from municipal water systems and limiting regulation of agricultural runoff.[14] In addition, exemptions were granted to the oil, mining, and timber industries, ensuring that they would not be subject to the permitting process. However, leaders of the new Congress were unable to make the wholesale changes to the Clean Water Act that they wanted, because of veto threats and opposition from moderate members of Congress.

Reinventing Environmental Regulation

A significant legacy of the Clinton years is a change in the nature of environmental rulemaking, particularly in the area of water pollution. Major administrative reforms were accomplished in the 1990s because of presidential leadership. Clinton took the reins of his administration and changed how policy would be made and implemented in the future, though these changes would later be challenged. The nature of intergovernmental relations and centralized decision making was to change during the Clinton administration, in part due to what happened with water regulation and review.

The centerpiece of the Clinton Administration's reform agenda, the National Performance Review (NPR) encouraged greater reliance on voluntary, cooperative and market-based policies at the EPA. The review occurred at a time when critics both inside and outside government suggested that environmental policymaking be decentralized in favor of collaborative strategies that might better resolve disputes at the local and regional level. As Myrick Freeman concluded, progress on water pollution was excruciatingly slow, because existing policies were inefficient and more cost-effective methods of reducing water pollution were needed, including economic incentives (such as charges and marketable discharge permits).[15] Three trends were apparent by the late 1980s, all pointing to the futility of local land-use planning. First, centralized decision making resulted in uniform rules that did not reflect local concerns. Second, the top-down nature of the process provoked local resentment and resistance. Finally, the rigid nature of these policies had a low compliance rate and offered discouraged innovation on the part of regulators and regulated entities.[16] The assumption was that collaborative processes result in more durable and effective policies and the time to formulate such policies would be reduced. Moreover, as Norman Vig suggests, administrative reform, rather than any single policy or program, was perhaps Clinton's most enduring environmental accomplishment.[17]

Expanding regulatory negotiation was a key component of Clinton's administrative reform agenda, not unlike the negotiations over national forest management described in Chapter 3. Advocates of regulatory negotiation had long argued that principles of alternative dispute resolution could be used to formulate better rules by involving stakeholders earlier in the process.[18] There were few experiments with regulatory negotiation prior to the Clinton years, and Congress passed the Negotiated Rulemaking Act and Alternative Dispute Resolution Act in 1990 to encourage less adversarial rulemaking. Well before the NPR was complete, Clinton issued Executive Order 12,866 encouraging greater use of regulatory negotiation, even directing agencies to find at least one opportunity to conduct a negotiated rulemaking.[19] The Clinton administration and agencies hoped that conflict might be reduced and compliance increased. Clinton's initiative attempted to institutionalize negotiated rulemaking, and the technique was used to set standards for previously unregulated drinking water contaminants, cryptosporidium for instance, as well as to set allowable levels for the chemical disinfectants used to control these contaminants.

In 1995, President Clinton and Vice President Gore released their report, "Reinventing Environmental Regulation," which included a detailed agenda for the EPA. It stressed flexible approaches over rigid rules, collaboration and consensus over adversarial decision making, and market incentives and performance standards over command-and-control regulation.[20] The EPA led the federal government in adopting of Clinton's vision, beginning with the Common Sense Initiative (CSI) and Project XL (Excellence in Leadership). CSI created an alternative to command-and control regulation for entire economic sectors by allowing them to negotiate performance-based environmental management standards industry-wide. CSI was the only example of such a partnership already underway when the Democrats lost their Congressional majorities in 1994.[21] The idea for Project XL, which began in 1995, was provided by the Aspen Institute's "Alternative Path" initiative.[22] Companies were encouraged to negotiate with regulators and develop site-specific pollution control plans, and command-and-control standards would be waived if plans could deliver superior performance. Three early experiments in collaborative environmental management were already underway before Clinton took office: the EPA's Chesapeake Bay Program, the Great Lakes Program, and the National Estuary Program.[23] Expanding upon these initiatives and introducing new partnerships were central to his plan to reinvent environmental regulation. Consequently, by 2008, the EPA and 20 states established such collaborative leadership programs.[24]

Ecosystem-Based Management (EBM) is another result of the Clinton Administration's effort to rethink its regulatory approach. As a holistic alternative to the existing piecemeal regulatory structures, EBM was introduced to address nonpoint as well as point-source pollution. States were held responsible for developing plans for the nation's 2,110 watersheds.[25] The approach became better known as watershed management, as administration officials cultivated support for the initiative with Congress and local governments. According to Michael Dombeck, who served as acting BLM Director early in the Clinton Administration, "We began talking about watershed management instead of ecosystem management, because everyone knows what a watershed is."[26] Ecosystem-based management was credited by the Clinton Administration for reducing pollution and political conflict. However, the link between the collaborative nature of EBM policies and the improved environmental conditions has been called into question, since much of the progress could have been a function of increased funding and the passage of time.[27]

Congress became a more receptive forum for opponents of conservation and pollution control policies in the 1990s at roughly the same time the executive branch became a stronger advocate. Divided government and a political stalemate within Congress prevented major policy change through the legislative process. As this gridlock intensified, Congress developed a preference for using the appropriations process to influence environmental policy.[28] Among the examples of this strategy were the many legislative riders concerning natural resources and energy Congress passed during the Clinton, Bush, and Obama years (see Chapter 3). At the other end of this gridlocked landscape, presidents often advanced their own environmental agendas without the cooperation of Congress. The following section examines current challenges in water quality as well as the executive-centered policy responses.

Nonpoint-source Pollution

By the time Congress amended the Clean Water Act in 1987, it was apparent that existing programs failed to adequately address the growing problem of nonpoint-source (NPS) pollution, especially urban and agricultural runoff. Urban runoff has become a significant problem because sprawl has claimed a larger area of previously undeveloped land. When porous terrain is paved over, there is an increase in both the volume of urban runoff and the pollutant load. Untreated runoff includes oil and toxic chemicals from automobiles, fertilizers and chemicals from gardening, as well as viruses, pathogens and heavy metals. When it reaches streams, it poisons fish and native vegetation, fouls drinking water supplies and makes recreation areas unsafe.[29] During periods of heavy rainfall and snowmelt, municipal sewers can become overwhelmed, with the excess volume discharged directly into rivers and streams. Stormwater discharges, including industrial was-tewater overflows, are considered point-sources and are easier to monitor and manage. Other sources present a greater challenge. Agricultural runoff from farms and animal feedlots have increased the pollution load in rivers and streams and contributed to eutrophication in lakes and estuaries.

Nonpoint sources were not addressed adequately in the 1972 law, according to Daniel Press, was written by Congress "to address what they could see, namely municipal and industrial sewage outfalls."[30] The 1987 law authorized the EPA to establish a NPS management program. This became the basis for requirements that the states develop their own management plans under the CWA, and the Coastal Zone Management Act, in the case of coastal watersheds. Although Congress did not specify the method of pollution control, the law established an EPA role in planning, funding and implementation of state programs. The act also marked a change in the relationship between the federal government and the states over water policy. According to John Hoornbeek, the law indicated that Congress was less willing to prescribe the means of pollution control or allow the direct regulation by the EPA.[31]

Even as the states complied with the planning process during the 1990s, many lacked the monitoring and enforcement mechanisms to control the problem. The Clinton Administration moved to use its existing authority to control pollution from nonpoint sources more aggressively. The administration unveiled a Clean Water Action Plan in 1998 that identified strict regulation of total maximum daily loads (TMDLs) as the most promising tool to control NPS pollution. TMDLs are the ambient levels of pollution in waterways, taking into account nonpoint and point-source discharges. The EPA eventually wrote TMDL standards and published a draft rule in 2000 in response to a court order.[32] The rule would have required states to set pollution limits for more than 20,000 waterways and require certain industries, including agriculture, to apply for federal permits for the first time. Through a leg-islative rider, Congress prohibited the EPA from spending federal funds to administer the rule and asked the National Research Council to evaluate the rule. When the NRC returned with a critical report in 2001, President Bush took the opportunity to postpone the rule, withdrawing it altogether in 2003.

Although no new environmental legislation was enacted during the entire two-term Bush/Cheney presidency, there were significant policy changes as a consequence of their administrative strategy. The administration was more willing to use the administrative process to advance its antiregulatory agenda on the environment than pursue policy change in the form of legislation.[33] If Congress and the administration rarely clashed over water policy during this administration, in sharp contrast to previous administrations, it is largely because the Republican leadership shared their goals. One noteworthy act of Congress was the Energy Policy Act of 2005, which amended the CWA to exempt oil and gas fields from permitting requirements under the National Pollutant Discharge Elimination System (NPDES). Because energy development was a top priority for the Bush Administration, the environmental agenda of his two terms was an administrative strategy to remove roadblocks for regulated industries, especially oil, gas and mining, as discussed in detail in other chapters.

Strip Mine Debris

Another action that had a significant impact on water quality was a strip-mining rule that reclassified mine waste as "fill." When the practice of mountaintop removal mining began, coal operators would often deposit debris into Appalachian rivers and streams, in violation of the Clean Water Act as well as the Surface Mining Act, which prohibits mining within 100 feet of a stream.[34] These valley fills tend to stay indefinitely, except when floods and dam collapses send rock and mudslides rushing into Appalachian waterways. The practice had been allowed because the Interior Department's Office of Surface Mining interpreted its own rules to apply only to the extraction of coal, not the waste produced from the blast. In 1999, a US District Court sided with Appalachian community activists in a lawsuit, which demanded that the Buffer Zone Rule also be enforced upon debris from mountaintop removal mining. Two years later, the coal industry sought relief from Vice President Cheney's Energy Task Force, which in 2001 recommended a clarification of the rule that the industry would find more favorable. After the first attempt to rewrite the Buffer Zone Rule was struck down as too vague, the Interior Department's Office of Surface Mining proposed a new one in 2005 that prohibited mining within 100 feet of streams, while explicitly exempting valley fills from the buffer zone.

The replacement Buffer Zone rule took effect in December 2008, early enough to prevent the incoming Obama Administration from reversing the policy without undertaking a costly and time-consuming new rulemaking.[35] Interior Secretary Ken Salazar announced that repealing and replacing the Bush-era rule was at the top of the department's agenda. The 2008 rule was struck down by a federal court[36] eventually, but the new rulemaking process was already underway. A 2009 Memorandum of Understanding between the Interior Department, EPA, and Army Corps of Engineers determined that a series of interagency actions were needed to reduce the harmful environmental impacts of surface mining.[37] The subsequent Environmental Impact Statement, rulemaking and extended comment period leading to the final rule would last for the duration of Obama's two terms. The moved at a deliberate pace in order to ensure all sides would have input, and that the rule would be able to withstand legal challenge.

The final Stream Protection Rule took effect at the end of 2016. In their determination to "get it right," regulators exposed their work to a different kind of risk: the legislative veto. The Congressional Review Act (CRA) had been invoked to overturn an administrative rule only once before, but the surprise victory of Donald Trump along with Republican majorities in both houses of Congress created the perfect confluence of circumstances to use it again. Trump had promised to use the power of the federal government to prop up the coal industry. One of the first actions of the 115th Congress was a resolution to repeal the Stream Protection Rule, and under the CRA, the bill could not be filibustered in the Senate. In an Oval Office ceremony, Trump proudly held up the page bearing his signature for the television cameras.

The Shrinking Scope of the Clean Water Act

At a 2016 campaign rally in Des Moines, then-candidate Donald Trump pledged to "eliminate job-killing regulations like the Waters of the U.S. Rule," to the delight of a cheering crowd. Though well-received, the remark seemed out of place in an otherwise generic stump speech about Americans losing jobs to illegal immigrants, the threat posed by Syrian refugees, and crime in Chicago. Most of the speech concerned his opponent's e-mails and private server.[38] Apart from his mention of the 2015 Waters of the US (WOTUS) rule, the speech was short on issues of concern to Iowa farmers. It is possible that Trump, the owner of several golf resorts, was already familiar with the issue because the WOTUS rule was opposed by the national association which represents golf courses. Nevertheless, he understood the rule was unpopular in Midwestern farming communities, promising to reverse the rule if he became president.

The pledge was the latest twist in an ongoing battle over the scope of the Clean Water Act that began with a pair of Supreme Court decisions in 2001 and 2006. In those cases, the federal government's jurisdiction over wetlands was challenged on the basis of language in the 35-year-old law referring to "navigable" waters. The largest setback occurred in 2006 when the court issued a decision in *Rapanos v. United States* invalidating the EPA's application of the NPDES requirement to most of the nation's streams, without articulating a new standard.[39] Because the rulings appeared to limit the number of waterways in which the EPA was authorized to control pollution, enforcement declined precipitously during the Bush/Cheney years.

The Bush Administration responded to the rulings with policy directives in 2003 and 2007, instructing the EPA and U.S. Army Corps of Engineers to avoid enforcement actions over small streams and other waterways, where jurisdiction was unclear and might have to be proven.[40] Consequently, hundreds of cases were suspended during Bush's second term. There were more than 2,600 enforcement actions in 2001, a figure that dropped annually until 2008, a year with less than 1,600 actions. Over the same eight-year period, the overall number of violations climbed from fewer than 19,000 to more than 27,000 in 2008.[41] Ohio, second only to Pennsylvania in the number of permitted facilities, had the poorest enforcement rate. Ohio had one of the highest violation rates, yet state officials took action against less than one percent of violators. One EPA official blamed the lack of enforcement for the increase in violations: "When companies figure out the cops can't operate, they start remembering how much cheaper it is to just dump stuff in a nearby creek." Another former Bush Administration official put it more bluntly: "Cases are now being lost because the company is discharging into a stream that flows into a river rather than the river itself," explained David Uhlmann, who prosecuted environmental crimes for the Justice Department.[42]

The effect of this policy change was to impair the CWA's ability to reduce the amount of nutrients and other chemicals from agricultural runoff. This runoff accumulates in larger bodies of water, such as the Chesapeake Bay, and can contaminate sources of drinking water for over 100 million people.[43] The CWA's jurisdiction could be restored by enacting clarifying legislation, but thus far, Congress has been unable to do so. The closest it came was in 2009, when the Senate approved the Clean Water Restoration Act, which would have removed the problematic word "navigable" from the text of the original 1972 law. The wording change was never considered in the House due to intense lobbying from the American Farm Bureau Federation and the National Association of Home Builders.[44] After the bill died in the House, President Obama sought to resolve the jurisdictional issues that arose during the previous decade. He directed EPA and Army Corps of Engineers to promulgate rules defining which wetlands and streams were covered by the law.

The WOTUS rule, finalized in 2015 after a lengthy comment period, was immediately controversial. The American Farm Bureau Federation mobilized opposition with the aid of a social media campaign. In addition to stories on Facebook, opponents used a Twitter hashtag, #DitchTheRule, to spread rumors that farmers would soon be required to seek permits for puddles and ditches. The EPA countered with its own social media campaign and Twitter hashtag, #DitchTheMyth, to counter misinformation about the rule, going so far as to coordinate with environmental and liberal advocacy groups to flood the Army Corps of Engineers and EPA dockets with supportive comments.[45] The final rule was challenged in federal court by 27 states. Upon joining the complaint, South Carolina Attorney General Alan Wilson claimed, "The EPA's proposed expansion would bring many roadside ditches, small ponds on family farms, water features on golf courses, and storm water systems under extremely burdensome federal regulation."[46]

The WOTUS rule was delayed due to a stay by the Sixth Circuit Court of Appeals, but the Trump Administration was unable to simply roll it back like so many other regulations. In his second month in office, Trump signed an executive order directing EPA Administrator Pruitt to undertake a new rulemaking process to replace WOTUS, a process that normally takes years.[47] In the meantime, the stringency of point-source enforcement continued to be determined by the states.

Factory Farms

Though a major point source of water pollution, most livestock feeding operations were still uncontrolled in the first decade of the 21st century.[48] A concentrated animal feeding operation (CAFO) is defined as a facility where livestock is confined for at least 45 days per year, and food is brought to the animals, rather than allowing the animals to graze. These include dairies, feedlots and poultry farms. Annually, the 240,000 largest CAFOs generate 500 million of tons of animal waste, which threaten to pollute waterways and groundwater. Since most CAFOs lacked effective stormwater controls, their waste ponds would often overflow, creating unsafe levels of nutrients and pathogens in rivers and streams. This lack of stormwater control was the reason for a public health catastrophe in Milwaukee, where heavy rains caused a deadly cryptosporidium outbreak in the city's drinking water supply 1993.

The EPA had the authority to regulate runoff pollution from CAFOs under the CWA, but no serious effort was made until a 1992 court order. The consent decree between the Natural Resources Defense Council and EPA required the agency to issue rules within 10 years. As part of President Clinton's Clean Water Action Plan, the USDA and EPA were directed to produce a unified national strategy to reduce water pollution from animal feeding operations. In 1998, 11 "listening sessions" were held around the nation to solicit feedback on a draft strategy, leading to a proposed CAFO rule in the final month of Clinton's presidency. The rule expanded the number of feeding operations required to apply for permits, but small and medium-sized facilities were subject to voluntary guidelines. The proposal included new regulations for the practice of spreading manure on agricultural fields, as well.

After taking office in January 2001, the incoming Bush Administration suspended one midnight rule after another. The CAFO rule was ultimately withdrawn, and the administration began to develop its own version of the rule after consulting with the interests the rejected rule would have regulated. In November 2001, OIRA Director John Graham held a closed-door meeting with industry representatives, including lobbyists for the Cattlemen's Beef Association, the National Turkey Federation, United Egg Producers, National Milk Producers' Federation and the National Pork Producers' Council. The meeting was also attended by one representative of the EPA's Office of Water.[49] After a year, the EPA proposed a much more industry-friendly rule.

The CAFO rule, which took effect in 2003, required 9,000 of the largest animal feeding operations to apply for permits under the CWA. Environmentalists charged that the rule included significant loopholes by imposing the permit requirement only to facilities with actual discharges, exempting those whose discharges were caused by stormwater. Instead, large feeding operations were required to develop management plans to prevent overflow. The Bush Administration rule did not control runoff from animal waste used as fertilizer, nor did it require cleanup of polluted groundwater, preferring instead to leave those issues to the discretion of the states. The NRDC responded by filing a new lawsuit against the EPA, more than a decade after the CAFO controversy appeared to be resolved.

Portions of the 2003 rule were soon vacated, forcing the EPA back to the drawing board to develop new requirements that CAFOs to submit nutrient management plans.[50] For environmentalists, it represented a hollow victory. In its 2005 decision, the Second U.S. Circuit Court let stand the Bush Administration's waiver for CAFOs that did not plan to discharge waste directly into waterways. Therefore, the EPA issued a revised rule in November 2008, that exempted most CAFO's from the permitting requirement altogether. Permits were required only from those feeding operations that discharged directly into waterways, as with the 2003 rule, although they had to seek certification as non-discharging facilities. Soon afterward, members of the House Judiciary Committee on Capitol Hill heard from Robert F. Kennedy, Jr., the attorney for the group that sued the EPA over the previous rule. Kennedy charged that the Bush administration used the rulemaking process to undermine the goals of the CWA, and avoided the opportunity to protect water quality when ordered by the court in 2005:

> The EPA should have used its ample authority and discretion to assemble all the evidence available to it, collect further data, and determine that all large CAFOs discharge, based on the nature of their design and method of operation, or that at least some set of large CAFOs, those in floodplains or areas with sandy soils, or high water tables discharge because of their location.[51]

Though weaker than the original CAFO rule, the 2008 rule was challenged by the National Pork Producers and vacated by the Fifth U.S. Circuit Court in 2011. The court struck down the requirement that feeding operations apply for permits if they only proposed to discharge into waterways, as opposed to actual discharges.[52] In a separate action, environmental groups challenged the 2008 rule to demand technology-forcing standards, but this was settled with the Obama Administration. Under the terms, the EPA agreed to promulgate a reporting rule that required all large and medium-sized CAFO's to collect and submit data regarding waste volume and pollutant loads. Subsequently, the EPA proposed new rules to set national standards for CAFO discharges and extend permitting requirements to smaller facilities.

As with the NPDES example, policy would be driven by executive initiative in the face of fierce resistance from Congress. Even though the EPA moved cautiously, there was a swift backlash from the livestock industry and Congressional allies. During the summer before the 2012 election, the proposed permitting and data collection rules were quietly withdrawn by the EPA. Environmentalists regrouped and resumed their strategy of challenging CAFO discharges at the state level. When the Natural Resources Defense Council and Pew Charitable Trusts used the Freedom of Information Act to obtain facility-level data originally collected under the suspended reporting program, industry groups howled. "It is beyond comprehension to me that with threats to my family from harassment atop biosecurity concerns, the EPA would gather this information only to release it to these groups," protested a former president of the National Cattlemen's Beef Association. "The only thing it doesn't do it chauffer those extremists to my house."[53] Eight U.S. Senators from the Environment and Public Works Committee sent an enraged letter to the acting administrator of the EPA (Jackson's successor was still awaiting confirmation by their committee). Before the 2013 Farm Bill stalled, the Senate and House attached amendments to their versions prohibiting the EPA from releasing the names, physical addresses or coordinates of livestock operations.[54]

Stormwater

A major source of water pollution that remained to be addressed in the 1990s was stormwater runoff. Previously, many of these sources had no controls, and sprawling development exacerbated the problem. Congress had offered little direction, but the EPA developed regulations under the authority of the 1987 Water Quality Act, which allows stormwater to be regulated under NPDES.

In December 1999, the EPA promulgated a Phase II rule requiring smaller urbanized municipalities with populations of fewer than 100,000 people to develop stormwater plans.[55] The Phase II rule was a prime example of the administration delivering on its promise of collaborative approaches to emerging problems. The EPA and the states were to develop an improved "toolbox" of strategies, from which the local governments could select and adapt to revise their own stormwater plans. The agency consulted with stakeholders in local government and industry to approve stormwater management plans. Three stormwater sources are subject to NPDES: construction runoff, industrial sources, and municipal storm sewer systems (MS4). Most states are now authorized to implement the NPDES stormwater program. Every state, except Idaho, New Mexico, New Hampshire, Massachusetts and the District of Columbia was managing its own program by 2011.[56]

Aging sewers contributed to the problem of stormwater pollution. Blockages and leaks in the nation's neglected sewers contributed to approximately 40,000 spills of raw or inadequately treated sewage. Many local officials ignored the problem for so long because "sewer pipes are something literally out of sight, out of mind," as one EPA official said.[57] For example, one of the worst sewer systems during this period was in Orange County, California. From 1996 to 2000, more than 57 million gallons leaked countywide in 1,108 separate events, contributing to 109 beach closures.[58] The EPA began working with a federal advisory committee in 1996 to design regulations to reduce the portion of water pollution caused by such overflowing sewers.

In the week before Memorial Day, 1999, Clinton identified the sanitary sewer overflow rule as one of his administration's priorities, calling it vital to protect water quality at the nation's beaches. During Clinton's final month in office, EPA Assistant Administrator for Water, Chuck Fox, announced a proposed rule requiring permits for municipal sewers. Under the proposal, 19,000 cities and sewer districts were required to make upgrades and then be subject to rigorous cleaning and inspection standards. Among them were previously unregulated communities that relied upon regional treatment facilities because they had no treatment plants of their own. The incoming Bush Administration immediately suspended the proposed rule in 2001, and later proposed a rule of its own, which would have allowed additional discharges of untreated sewage into waterways. In 2003, the EPA proposed to allow so-called sewage "blending." This would have allowed treatment facilities to skip secondary treatment when confronted with stormwater overflows, thereby diverting filtered, but untreated, sewage from chemical treatment ponds and combining it with treated water, before discharging the mixture into waterways. In the face bipartisan opposition in Congress, the EPA withdrew the proposal in 2005.[59]

In addition to the sewer regulations, the Bush Administration used the budget as another powerful instrument to change the course of water pollution control. In 2002, the White House proposed a federal budget that reduced funding for projects to upgrade sewage treatment systems and control stormwater from its annual sum of $1.35 billion to $1.21 billion.[60] In response, the House of Representatives considered a bipartisan bill to set aside $20 billion over five years to help local communities pay for those sewer upgrades. Appearing on Capitol Hill, the head of the EPA's Office of Water, Benjamin Grumbles, warned that the White House would oppose the increased funding for clean water projects: "The president clearly defined his priorities in the State of the Union as defense and homeland security," he testified to the House Committee on Transportation and Infrastructure.[61] Later, the 2007–08 budget trimmed another $395 billion from the CWA revolving fund.

At the end of the decade, in absence of new legislation, the Obama Administration moved to address the health and environmental risks from storm runoff with the tools available. In 2011, EPA Administrator Lisa Jackson announced plans for a permit system for runoff under the EPA's existing statutory authority. Obama took another significant action to control storm runoff early in his presidency when he asserted presidential leadership to jump-start the cleanup of the Chesapeake Bay.

Healing the Chesapeake Bay

The example of Chesapeake Bay illustrates of the limits of the policy regime created in the 1970s and the importance of presidential leadership. This endangered ecosystem has been a laboratory for cooperative as well as coercive experiments in environmental management, with limited success. Despite a reduction in pollution from point-source discharges in the first decade of the NPDES, the overall health of the nation's largest estuary took a turn for the worst. Silt from coastal development and dredging, urban runoff from inland development, and nutrients from agricultural runoff and septic tanks caused further deterioration. There was a push for regional management, beginning with the founding of the Chesapeake Bay Commission in 1980. At President Reagan's direction, the EPA asserted a larger role with the Chesapeake Bay Program in 1983. Since the primary actors continued to be the states of Virginia, Maryland, and Pennsylvania and the District of Columbia, Reagan could use the bay as a symbol of his support for the environment while keeping his commitment to New Federalism.

As his reelection campaign approached, Reagan saw an opportunity to create an environmental legacy by declaring the bay a national treasure and asking the federal government to match state-level initiatives to clean up its polluted waters. After months of discussion among his domestic policy staff, Reagan decided to abandon the hands-off approach. In his 1984 State of the Union Address, he declared, "I have requested for EPA one of the largest percentage budget increases of any agency. We will begin the necessary effort to clean up a productive recreational and a special national resource—the Chesapeake Bay."[62] Reagan also supported a $52 million four-year cleanup effort, which Congress approved one month before the 1984 election.[63] There was little prospect that Reagan's election-year actions would affect his dismal standing with environmentalists. However, it was an important gesture in following the resignation of controversial appointee Anne Gorsuch.

A more comprehensive effort was launched in the last decade of the 20th century. For Clinton, the Chesapeake Bay represented the kind of problem that could be solved by the kinds of nonregulatory approaches that he championed. Consequently, during this period funding for scientific research was increased, environmental partnerships were strengthened, and market-based policies were introduced. In Virginia and Pennsylvania, marketable permits were introduced to control manure waste from dairy farms.[64] However, most of the state policies remained merely symbolic, such as voluntary programs to control agricultural runoff in Maryland and Virginia. The impact of this approach was minimal, because according to Layzer, state officials were reluctant to impose strict mandates because they were loath to antagonize farmers, increase ratepayers' fees, or curb lucrative development.[65]

When Clinton delivered his 1995 Earth Day address, he stood at the shore of the Chesapeake Bay to further position himself as the defender of the environmental policy regime that he had previously accused Congress of dismantling. He showcased the bay as the symbol of 25 years of progress, declaring "...all you have to do is look at this bay. The beauty here is God-given, but it was defended and restored by human beings. Not long ago, garbage floated on the Chesapeake; shellfish were unsafe to eat."[66] Beneath the seemingly pristine surface, conditions were actually worsening. Nitrogen and phosphorous levels began to increase, undoing the progress that was achieved by treating sewage, limiting point-source discharges and banning certain detergents. Low-oxygen "dead zones" appeared with greater frequency.[67] The problem could be observed in the region's growth, with a population of 13 million in 1990 surpassing 15 million by the end of the decade. Farmland, green spaces, and forests near the bay were lost and replaced with new pollution sources: suburban houses, streets, and parking lots.

Underscoring the problem was the disconnect between what people believed were the major sources of pollution and the actual sources of pollution. A survey of watershed residents found that 87 percent of people were either "very concerned" or "somewhat concerned" about bay pollution, and 61 percent agreed that more should be done to clean up the bay. The same poll found that residents of the Chesapeake Bay region did not see themselves as responsible for the problem. They were more likely to blame industrial sources, such as factories, despite significant reductions in effluent from those sources during the previous two decades. Only 10 percent identified population growth or individuals as major contributors to the problem.[68]

As Howard Ernst explains, the problem was that Congress had created a bureaucracy but deprived it of the resources it needed to succeed.[69] For more than two decades, the commission followed its mandate to monitor and reduce certain pollutants deposited into the watershed. This piecemeal approach allowed the commission to meet easily achievable benchmarks. However, most of this improvement was indicated by measures of effort, rather than indicators that would gauge overall progress. Environmental scientists and policymakers lacked data regarding the overall health of the ecosystem, relying instead upon information on individual programs and indicators. The commission's own data reinforced the illusion of progress even as the quality of the bay deteriorated. This was not purely by accident, however. Regulators complained that they were under pressure to show results in order to protect Congressional appropriations for the project.[70]

There were no new federal initiatives during the Bush years, and the 43rd president's only public statement about the Bay came in 2007. In a weekly radio address, on his way to a fishing trip there, Bush pledged to clean up the Chesapeake Bay, " ... that our sportsmen depend on."[71] This symbolic statement of support reflected the perspective of some hook-and-bullet groups that had some access to his administration, as opposed to the mainstream environmental community, which valued the bay for more than its recreational uses.

As the Bush years came to a close, a group of scientists alarmed about the deteriorating state of the bay prepared a report that they hoped would force the new administration to begin with a fresh start. On December 3, 2008, they assembled in Annapolis to declare that voluntary and collaborative policies had not only failed to improve the bay, but conditions had steadily deteriorated over the previous 25 years. Their report declared these incremental approaches a failure, calling for a federal policy devoted to overall quality rather than piecemeal indicators. At the same time, cash-strapped states of Maryland and Virginia were making deep cuts to their conservation and bay restoration programs. Even though some of these reductions were offset by increases in federal spending, there was a discouraging lack of forward progress.

Obama signaled a renewed commitment in 2009 by appointing former EPA Assistant Administrator Chuck Fox to a new position as senior advisor for the Chesapeake. In 2010, Obama moved to break the political deadlock. He then issued Executive Order 13508, reaffirming the estuary as a national treasure and asserted a leading role for the federal government. He named then-Administrator Lisa Jackson to head a new Federal Leadership Committee, which had the responsibility to overhaul the cleanup and issue annual progress reports. Although the order authorized no new regulatory powers or requirements for the states,[72] Obama followed up with a White House budget request for 2010 increased funding for the Chesapeake Bay Program to a record $35.1 million. In addition, the 2009 stimulus bill provided $878 million for the CWA revolving loan fund in Chesapeake Bay watershed states.

In a more substantial agency move, the EPA quickly went to work promulgating TMDL standards for the watershed. The final regulations, listed in January 2011, allocate the pollution load along the rivers, tributaries and tidal basins that comprise the system. The plan requires a reduction of 63 million pounds of nitrogen and 4 million pounds of phosphorous by 2025, in order to restore habitat for native fish, crabs, and waterfowl. States must develop watershed implementation plans and show how they will meet two-year benchmarks. Significantly, the new regulations are enforceable. If states fail to make progress, the EPA can impose water quality standards on upstream tributaries or require permits for additional sources such as wastewater treatment plant discharges and animal feeding operations.

Drinking Water

In 2015, the year unsafe levels of lead were discovered in Flint's tap water, there were approximately 18 million people nationwide who purchased their water from systems that violated the EPA's Lead and Copper Rule. Altogether, there were 18,000 community water systems with at least one violation of the Safe Drinking Water Act. These include violations of health standards, violations of testing requirements, and failures to report violations of health standards.[73] The chemical and microbial contaminants found in drinking water are largely invisible, and unlike air quality, the quality of drinking water is often unknown to the public. Since few citizens have the expertise to assess the health risk from their drinking water on their own, the debate relies on competing claims by scientists, regulators and interest groups. This is especially confounding when it comes to natural contaminants such as arsenic and radon. When people consider the quality of their drinking water, they tend to think about industrial and organic pollutants. As difficult as they are to detect, pollutants such as lead and mercury are commonly understood to be the byproducts of industrial pollution and improperly treated wastewater.

A decade after the SDWA was enacted, Congress was frustrated with the EPA's plodding pace and amended the law in 1986 to include hammer clauses requiring that the agency issue or revise standards for 83 specific contaminants by 1989, and another 25 new contaminants every three years. The EPA missed its deadline for all but a handful of the required standards and lagged behind on rulemaking for many of the required pollutants, including chlorine disinfection byproducts, arsenic, and radon. Following the election of 1992, those standards still had not been approved, and Congress began to consider new drinking water legislation. When the 103rd Congress deadlocked over the new drinking water bill, it became one of the initiatives that fell by the wayside before the 1994 election.

When the SDWA was amended again in 1996, the law reflected priorities of the new Republican majority as well as President Clinton. The 1996 SDWA Amendments featured innovative provisions, including the use of information as a regulatory tool. Clinton became an advocate for a right-to-know provision, modeled after the Toxic Release Inventory (see Chapter 5). It required water systems to send their customers annual reports listing what pollutants were found, their potential health effects, and whether any federal water standards were violated. The provision had been stripped from a senate version of the bill in 1995 after heavy opposition from water companies, state and local governments, but restored when the bill was reintroduced.[74] Clinton also supported a requirement in the bill that residents receive 24-hour notice when dangerous contaminants are discovered, and another provision creating an online database of water quality data for all 2,110 watersheds, searchable by zip code ("Surf your shed"). The legislation also created a revolving loan fund, modeled after the Clean Water State Revolving Fund, in order to upgrade community water systems.

Congress rescinded the requirement that 25 new contaminants be regulated every three years, replacing it with a requirement that setting limits for new pollutants must offer a "meaningful opportunity for health risk reduction." The new language also required the EPA to show that the new standards are financially justified, a mandate that has impeded subsequent efforts to set standards for new drinking water contaminants. In 2016, the EPA issued a health advisory for PFOA and PFOS, chemicals used to manufacture consumer products including Teflon and Scotchguard after they were linked to cases of cancer and birth defects along the Ohio River in West Virginia. The advisory recommended a voluntary lifetime limit of 70 parts per trillion for both chemicals, but the Trump EPA under Scott Pruitt decided against setting standards, citing the cost to the chemical industry.

Microbial Contaminants and Disinfection Byproducts

The EPA's Office of Water began a negotiated rulemaking in 1992 to set standards for previously unregulated pathogens in drinking water systems. At the time, the agency estimated that microbial contamination was the cause of 79 percent of Maximum Contaminant Level (MCL) violations. A 1989 lawsuit by an Oregon group, the Bull Run Coalition, held the agency to task for missing the statutory deadlines of the SDWA Amendments. The EPA in turn invoked the Negotiated Rulemaking Act of 1990 and began the process of collaborating with stakeholders to develop standards for microbial contaminants. Negotiations were held with representatives from regulated industries, including water utilities, manufacturers of chemicals used for water treatment, environmentalists and public health groups. The negotiated rulemaking was an attempt to set standards that would be acceptable to the municipalities and the other interested parties who were represented at the table.

Shortly after negotiations stalled in 1993, the city of Milwaukee experienced the outbreak of Cryptosporidium in its water supply that killed over 100 people and made thousands of others sick. Annually, cryptosporidium is responsible for an estimated 573,000 illnesses, and the high concentration of victims and large death toll raised the profile of this pathogen considerably at a time when the agency began serious discussions about revising the standards.[75] Clinton invoked the Milwaukee tragedy, which occurred within weeks of his inauguration, when he proposed to update and strengthen the SDWA later that year. However, the drinking water bill became a casualty of Congressional gridlock, and was abandoned before the 1994 election, along with proposed amendments to the CWA. However, the cryptosporidium deaths brought the parties back to the table, and the EPA was able to promulgate a Stage 1 enhanced surface treatment rule in 1998 that applied to community water systems serving over 10,000 residential units, which took effect in 2002. The Stage 2 rule, which applies to community water systems serving at least 25 residents nationwide, appeared in 2005.

The pressure to eliminate microbial contaminants raised the risk of another public health problem. Several of the 83 contaminants Congress identified in the 1986 SDWA Amendments were waterborne pathogens that were usually controlled by the widespread use of chemical disinfectants. Except for a chlorine standard issued in 1979, the EPA did little to control the health risks from chemical disinfectants and disinfection byproducts (DBPs) in drinking water because the risk of drinking untreated water was significantly greater. The EPA faced a dilemma because the SDWA required the agency to set Maximum Contaminant Level Goals (MCLGs) for any contaminant that may have an adverse effect on human health and is known to be present in public water supplies, including DBPs such as chloroform, which were suspected carcinogens. Finally, when the EPA began the negotiated rulemaking for the related problem of microbial contamination in 1992, the agency decided to set MCLGs for byproducts of those chemical disinfectants, such as chlorine, that posed significant health risks. The rulemakings for microbes and DBPs were undertaken simultaneously so that the agency could develop an approach that would best ensure a significant reduction in risk from DBPs without compromising the public's protection from microbes.

The rulemaking process was reportedly hindered by the high level of uncertainty associated with DBPs. The uncertainty is illustrated well by the controversy over the proposed chloroform standard. At the time of the DBP rule's proposal in 1994, the additional lifetime cancer risk level from chloroform was estimated at one in 10,000, assuming a concentration of chloroform of 600 microns per liter of drinking water. However, there was still significant uncertainty about the concentration of chloroform and other DBPs in public water systems, and about the effects. Because the nature of the carcinogenicity of chloroform was not as yet understood, regulators fell back on SDWA guidelines for carcinogens, which require a MCLG of zero when there is insufficient information on the toxicity of a substance. However, the guidelines allow the agency to set a MCLG above zero if the available science supports the existence of a safe threshold.

The use of regulatory negotiation did not help the EPA avoid one of the major pitfalls of traditional rulemaking: litigation. Despite their participation in the rulemaking, manufacturers of chemical disinfectants filed a lawsuit to block the standards. The manufacturers claimed that additional data on the health effects of chloroform became available after the rules were proposed in 1994. The new research demonstrated that a minimum level of chloroform was required for toxicity, suggesting a nonlinear dose-response relationship.[76] On the basis of this evidence, an alternate MCLG of 0.3 mg/L was proposed in a subsequent notice in the Federal Register, and comments were solicited for both the zero standard and the nonzero standard. Ultimately, the EPA retained the zero standard in the final rule, despite mounting evidence in support of a threshold. Manufacturers of chemical disinfectants accused the EPA of "sacrificing science at the altar of politics" by refusing to raise the standard.[77] In 2000, the US District Court in Washington, DC, vacated the zero standard for chloroform, ruling it was arbitrary and capricious because the agency had failed to use the best available science.[78]

The health risk from disinfectants is a countervailing risk to that of microbial contamination. Establishing limits on DBPs limits the range of options for water systems to control microbials. The failure to chlorinate drinking water threatens vulnerable subpopulations, including children, the elderly and those with weak immune systems.[79] Choosing to reduce the risk from one public health threat at the peril of increasing another risk presents a difficult tradeoff. There might have been standards for DBPs and microbial substances earlier, were it not for this dilemma.

Another provision of the 1996 SDWA directed the EPA to focus on regulating those pollutants that pose the greatest health risk. By the year 2000, however, the EPA still had not issued new standards for two of the most toxic drinking water contaminants, arsenic and radon. Only a weak standard for arsenic dating back to World War II was in place, and radon remained unregulated in drinking water. Eventually, standards for both contaminants were proposed, suspended and then re-proposed under politically charged circumstances.

Radon

The discovery of high radon concentrations in private homes made national headlines in the 1980s, but neither Congress nor the Reagan Administration responded to the issue with the same fervor that met other focusing events, such as Love Canal, a few years earlier. The incident led to widespread testing, which confirmed that radon contamination in homes was much more common than previously believed. However, the federal government's initial response was to avoid regulation. The Reagan Administration was staunchly opposed to giving the EPA broad regulatory power to control radon. Because most exposure occurred on private property—within the home—there was little public support for government invading the sanctity of that private space, and Congress was reluctant to attempt a command-and-control solution to a new and poorly-understood environmental problem than the administration. Initially, the federal response to the threat of radon dealt with airborne radon gas and was aimed at educating private citizens, who would have the burden of detecting and mitigating radon.

The administration's reluctance to impose any stronger regulations was undermined by tension within what Edelstein and Makofske term the "Ideology of Reaganomics," which stressed cost-benefit analysis as much as hostility to regulation. The OMB, which was extremely influential in the Reagan White House, identified radon regulation as a cost-effective way to save lives.[80] This view was supported by the EPA's comparative risk project that suggested the gas had been an overlooked health threat.[81]

The problem of radon in drinking water made it onto the EPA's regulatory agenda in the 1980s after Congress forced it there. The 1986 SDWA Amendments, in requiring that the EPA set health-based standards for substances found to pose a significant risk to human health, specifically mentioned radon. The Bull Run Coalition lawsuit was filed in 1989 over the agency's failure to issue final standards for radon (among other drinking water contaminants) as required by the 1986 SDWA Amendments. Even though the deadline was not met, it forced the agency to act quickly and propose a standard before it was ready to do so. The availability of new scientific information was cited as a secondary reason for radon's ascent to the regulatory agenda. Scientists were still learning about the occurrence of radon in ground water and public water systems when Congress announced a deadline for the EPA to set a standard. The problem was already high on the scientific community's agenda, and new research that would influence regulatory decisions was still being published.[82] For example, a 1994 revised EPA analysis found considerably higher risk from ingestion of radon in drinking water compared to the estimates used for the 1991 proposed rule.[83]

The first standard for radon was proposed in 1991, as part of a rule regulating radionuclides in community drinking water systems that use ground water. As a known carcinogen, radon's MCLG was automatically set at zero. Because drinking water contributes such a small part to the overall radon in the human environment, water utilities raised objections over the cost of the proposed rule. Under pressure from Congress, the EPA delayed the final rule, and it was ultimately withdrawn. After the SDWA was amended in 1996 to allow more flexible approaches to risk reduction, the EPA withdrew the standard and started anew by proposing an Alternate Maximum Contaminant Level Goal (AMCLG) for radon. States were permitted to create their own multimedia mitigation plans to regulate radon. With EPA approval, states could meet the AMCLG, in conjunction with other strategies to reduce radon in the air, as well as education programs.

Arsenic

Radon was not the only uncontrolled toxic substance in the nation's drinking water at the end of the 20th century. Long after the SDWA amendments mandated tougher standards for pollutants that presented the highest risk, By the year 2000, the EPA had missed three statutory deadlines to update a weak World War II era standard for arsenic. As with the proposed radon standards that were withdrawn during the Clinton years, a proposed arsenic standard was suspended by the Bush administration when the EPA was beset with criticism that the regulations were too costly.[84] There were further similarities. Both rules were to be imposed upon community water systems, which would be forced to bear the cost of mitigation. There also was a scientific consensus that both contaminants posed a very high risk to human health. The major difference between the cases was the public's reaction. When the EPA quietly withdrew the radon rule and began work on a more flexible rule, there was nothing approaching the public backlash to the arsenic decision.

By the time the SDWA Amendments of 1986 mandated stricter standards for certain contaminants, controversy over the drinking water standard for arsenic had existed for decades. A 50 ppb standard left over from World War II was deemed inadequate by environmentalists. The 1996 SDWA Amendments required the EPA to concentrate on those contaminants that posed the greatest risk to public health. The EPA responded in 2000 by proposing a 5 ppb standard, which water utilities complained was too strict and costly to implement.

Key opposition came from organizations charged with implementing the new standard, including the American Water Works Association a trade group with 57,000 members, the American Water Works Research Foundation, a think tank funded by water utilities, and the National Rural Water Association, which represented more than 20,000 small communities. To a lesser extent, there was some opposition from industries whose wastewater contributed to arsenic contamination, such as the mining and wood preserving industries. Instead, they argued for standard of 10 ppb, a level that would bring the US in line with other industrial nations. The World Health Organization had adopted 10 ppb as the maximum level for drinking water years earlier, as had the European Union and Japan.[85] A stricter standard, they argued, was too burdensome for small communities and that the additional health protection of a 5 ppb standard was not worth the cost. The Clinton Administration yielded to these arguments and approved the 10 ppb standard in a final rule issued during its last week in office.

Although interest group opponents had succeeded in persuading the Clinton administration to lower the proposed standard from 5 ppb to 10 ppb, they were elated by the Bush administration's decision to suspend the rule. Even prior to his inauguration, Bush announced his intent to review all federal regulations issued at the end of his predecessor's term. One of his first actions after assuming office was to suspend a handful of new regulations, including the new arsenic rule. Incoming EPA Administrator Christine Todd Whitman cited the need for more scientific research and a review of the rule's costs and benefits. Here, the review was a pretext for delaying and possibly rescinding the Clinton Administration's 11th-hour rule. The administration took political cover in an OMB report to Congress that highlighted letters from regulated entities that questioned whether the benefits of the arsenic standard justified the costs. The two letters were submitted in response to a call from OIRA Director John Graham for recommendations to improve regulation. One letter, authored by the Public Works Director of the City of Austin, Texas, suggested that the EPA overestimates benefits and underestimates costs. In the second letter, the American Water Works Association argued that the benefits of the 10ppb standard did not exceed its costs according to the EPA's own analysis.[86]

Environmentalists and an appalled public put the administration on the defensive over its decision, and a Republican-dominated Congress responded to by passing legislation directing the EPA to implement a strict arsenic standard. Although a scientific review vindicating the tougher standard was completed later that year, the opposition's success at delaying the rule by calling for more scientific research, or paralysis by analysis, was seen as an effective, yet cynical tactic.[87] Congress was under pressure from the public to impose the stricter standard through legislation, and even Congressional Republicans began calling for the rule's reinstatement. One Republican critic was Vermont Senator Jim Jeffords, who denounced the administration's environmental stances before defecting from the party to become an Independent. After the Republican-dominated House of Representatives passed a bill endorsing the standard, the administration backed down. On October 31, 2001, the EPA published a final rule that used the same arsenic standard proposed in January.

Ironically, a stricter arsenic regulation was issued after the Bush Administration's reversal. Because of a loophole in the Clinton Administration rule allowing 10 ppb to be measured as 0.01 mg/L, municipalities that detected arsenic levels up to 0.015 mg/L could easily comply by rounding their measurements down. Following the controversy, the EPA reinstated the standard and closed the loophole. In 2002, the EPA published a clarification to the revised final rule setting the standard at 0.010 mg/L.[88] For Bush, the controversy over the arsenic rule was an early political setback with no lasting impact. By the time it was resolved in October, 2001, the threat from arsenic in drinking water was no longer under close public scrutiny. Since the Clinton Administration would not be the one to enforce the rule, it might have intentionally passed this proverbial hot potato to the next administration. Had the Supreme Court ruled differently in Bush v. Gore, and the final arsenic rule issued on schedule, then Gore would have inherited the controversy.

In 2015, the EPA completed a rulemaking to control a significant source of arsenic and other heavy metals before they enter municipal water supplies. Coal-fired power plants had become the nation's leading source of toxic wastewater, and more than a third of coal plants discharge wastewater within five miles of a down-stream community's drinking water intake. The new rule required coal plants to demonstrate they were using the best-available technologies to remove lead and mercury as well as arsenic from their wastewater by 2018, but in March 2017, Scott Pruitt declared a stay of the deadline. Pruitt announced a reconsideration of the rule, arguing the EPA had underestimated the cost of compliance and overestimated its benefits.[89] As Chapter 5 will show, the Trump Administration repeatedly used this justification to rescind Obama-era regulations, often to the coal industry's benefit.

Conclusion

The case of water quality illustrates the transformation of the policy agenda in the decades since the first Earth Day. The salience of first-generation problems, the fouling of waterways with industrial effluent and raw sewage, moved the mass public to participate in protests and punish unresponsive legislators at the ballot box. Throughout the 1970s and 1980s, Congress as a whole was sensitive to the public's concerns. Major legislation to federalize water pollution policy was enac-ted, sometimes in the face of presidential opposition and vetoes. Congressional oversight forced presidents and the executive bureaucracy to implement the law.

The politics of water shifted in the 1990s. Historical public opinion data reviewed in this chapter show that water quality continued to be the top envir-onmental problem as perceived by the public, but environmental issues fell behind other priorities. This sea change was reflected in Congress, where there was no longer the same urgency to strengthen water quality. The water pollution challenges at the dawn of the 21st century were stormwater, agricultural runoff and urban runoff, issues that did not inspire the same mass outrage as the first-generation problems. To the average citizen, out of sight truly equaled out of mind, as many of the new threats remained just invisible, such as chemicals and pathogens in drinking water, and leaky and crumbling sewers. The public was disengaged, and accordingly Congress became indifferent, and even at times hostile, to new water quality challenges.

Again, and it becomes even more apparent in the case of water, the policy role of the presidency has evolved along with the politics of the environment. Beginning in the third decade, presidents have been the advocates who forced water pollution onto the agenda. When Congress took its time reauthorizing legislation in the 1990s, Bill Clinton spoke of the dangers of inaction by going public, reminding the country of the deadly consequences of microbial contamination of surface water. Consequently, there has been a role reversal between the branches of the federal government. Major policy change occurs not with legislation, but when the president becomes engaged and shows leadership. Barack Obama appointed water quality experts to key administrative positions and took administrative action to reorganize the Chesapeake Bay cleanup and intensify enforcement of NPDES.

The politics of the environment have changed since the first Earth Day because the scope of conflict has narrowed. Within government, advocates for policy solutions to the new problems were more often presidents and specialists within the bureaucracy. Bill Clinton used a variety of administrative tools to accomplish several important water quality goals, as well as alternative tools such as negotiated rulemaking, watershed management, and various partnerships, which originated with his reinvention initiatives. As with the discussion of natural resource management in Chapter 3, Clinton's willingness to support proposals from the career bureaucracy was the foundation of his administrative presidency, strengthening his position relative to Congress. This is an important reason Daynes and Sussman conclude about the Clinton years, "...his political strength and the leverage he had with Congress could be questioned and labeled insufficient.... Thus his environmental legacy often depended on finding ways to encourage policy acceptance while avoiding party confrontation within Congress."[90]

The presidency of George W. Bush featured reversal of water policy priorities, but a similar preference for administrative policymaking. For most of his presidency, Bush had the advantage of Republican Congressional majorities and Congressional leadership that was sympathetic to the administration's deregulatory agenda. Nonetheless, the Bush Administration used the rulemaking on numerous occasions to provide regulatory relief for key interests, notably agriculture, municipal water systems, and the mining, oil and gas industries. The Obama presidency represents a redirection toward more activist and collaborative water policies. As with the Clinton Administration, Obama's agenda emphasizes partnerships and collaborative policies. Often thwarted by opposition Republicans, there was little prospect that Obama would persuade Congress to enact significant environmental legislation in the new landscape. His administration pursued ENR priorities without Congress as much as possible, relying upon directives and administrative rulemaking to reverse Bush-era policies on water pollution. The example of the Chesapeake Bay demonstrates that he was willing to assert more federal control when confronted with mounting evidence that nonregulatory approaches proved insufficient.

Obama's legacy turned out to be fragile, as President Trump and the 115th Congress easily repealed his landmark Stream Protection Rule, and the Trump Administration took aim at rewriting the Waters of the United States Rule. A common thread in all four eras is that Congress was often passive as the executive branch initiated major changes to water policy. The case of water policy is not only illustrative of the changing policy landscape, it is also exceptional because pollution of waterways and drinking water are consistently ranked by voters in the top tier of environmental issues. It stands in contrast to other pollution problems which are less understood by Congress and the general public. The following chapter investigates the evolving role of the presidency in the problematic area of toxic and hazardous waste, an issue with a lower level of salience to the public. After Congress created the Toxic Release Inventory in the 1980s, the executive branch became the source of subsequent toxics policy innovation. This led to successful state-level recycling and brownfield reclamation programs during the Clinton years and the Bush/Cheney Administration's advocacy for voluntary hazardous waste policies and marketable permits for mercury emissions.

Notes

1 For a compelling first-person account of the Flint Water crisis, see Mona Hanna-Attisha, 2018, What the Eyes Don't See: A Story of Crisis, Resistance, and Hope in an American City, New York: One World.
2 Ibid.
3 Michael Wines and John Schwartz, 2016, "Unsafe Lead Levels in Tap Water Not Limited to Flint," New York Times (February 8)
4 A Reuters investigation later that year revealed more than 3,000 municipalities with similar or worse drinking water quality. See M. B. Pell and Joshua Schneyer, 2016, "The Thousands of U.S. Locales where Lead Poisoning is Worse than Flint" (December 19), www.reuters.com/investigates/special-report/usa-lead-testing/.
5 Former EPA Administrator William Ruckelshaus, quoted in Charles Duhigg, "Clean Water Laws Neglected, at a Cost," New York Times (September 13, 2009): A1.
6 USEPA, 2013, "National Rivers and Streams Assessment," EPA/841/D-13/001, February 28.
7 Benjamin Kline, 2011, First Along the River, 4th edition, Lanham, MD: Rowman and Littlefield.
8 The largest sources of sewage leaks were older suburbs of Cleveland, including Garfield Heights, Cleveland Heights and Maple Heights, which did not have their own water treatment plants. The proposed upgrades in Northeast Ohio were expected to cost $3.6 billion. See John C. Kuehner, "EPA-Fueled Upgrades for Sewers Could Cost Millions," Cleveland Plain-Dealer, January 16, 2001: 1B.
9 Denise Scheberle, 2004, "Devolution," in Environmental Governance Reconsidered, eds Robert Durant, Rosemary O'Leary and Daniel Fiorino, Cambridge, MA: MIT Press.
10 A. Myrick Freeman III, 2000, "Water Pollution Policy" in Public Policies for Environmental Protection, 2nd edition, edited by Paul R. Portney and Robert N. Stavins, Washington, DC: 203.
11 For a more elaborate and nuanced discussion of the reasons for the sudden about-face in Congress, see Paul C. Milazzo, 2006, Unlikely Environmentalists: Congress and Clean Water, 1945–1972, Lawrence, KS: University of Kansas Press.
12 Bill Clinton, "Remarks on the 25th Observance of Earth Day in Havre de Grace, Maryland," April 21, 1995, Public Papers of the President: William J. Clinton: 1995, Washington, DC, 1: 565.
13 Clinton was asked by a reporter whether he worried his voice would be heard, given Congressional dominance of the political debate since the Republican victory. He responded, "The Constitution gives me relevance. The power of our ideas gives me

relevance. The record we have built up over the last two years and the things we're trying to do to implement it give it relevance. The President is relevant here, especially an activist President." See "The President's News Conference," April 18, 1995, Public Papers of the President: William J. Clinton: 1995, Washington, DC. 1: 547.

14 Benjamin Kline, 2011, First Along the River, 4th edition, Lanham, MD: Rowman and Littlefield.

15 A. Myrick Freeman III, 2000, "Water Pollution Policy," in Public Policies for Environmental Protection, 2nd edition, edited by Paul R. Portney and Robert N. Stavins, Washington, DC: 203.

16 Ibid.

17 Norman J. Vig, 2006, "Presidential Leadership and the Environment" in Environmental Policy: New Directions for the Twenty-First Century, 6th edition, edited by Norman J. Vig and Michael E. Kraft, Washington, DC: Congressional Quarterly Press, 100–123.

18 For example, see Philip J. Harter, 1982, "Negotiating Regulations: A Cure for Malaise," Georgetown Law Journal 71(1): 1–118.

19 William J. Clinton, Executive Order 12866—Regulatory Planning and Review Online by Gerhard Peters and John T. Woolley, The American Presidency Project, www.presidency.ucsb.edu/node/227792, www.presidency.ucsb.edu/documents/executive-order-12866-regulatory-planning-and-review.

20 National Performance Review, 1995, "Creating a Government that Works Better and Costs Less," U.S. Government Printing Office.

21 Due to the 104th Congress' assault on environmental protection, subsequent reinvention initiatives rolled were often misconstrued as a defensive reaction to the Republican Agenda, as demonstrated by Daniel J. Fiorino. See Daniel J. Fiorino, 2006, The New Environmental Regulation, Cambridge, MA: MIT Press.

22 Various Project XL programs are described and evaluated in a volume by Alfred A. Marcus, Donald A. Geffen and Ken Sexton, 2002, Reinventing Environmental Regulation: Lessons from Project XL, Washington, DC: Resources for the Future.

23 John DeWitt, 2004, "Civic Environmentalism," in Environmental Governance Reconsidered, eds Robert Durant, Rosemary O'Leary and Daniel Fiorino, Cambridge, MA: MIT Press.

24 Jonathan C. Borck, Cary Coglianese and Jennifer Nash, 2008, "Evaluating the Social Effects of Environmental Leadership Programs," Working Paper of the Regulatory Policy Program, John F. Kennedy School of Government, Cambridge, MA: Harvard University.

25 Robert B. Keiter, 2003, Keeping Faith with Nature, New Haven: Yale University Press.

26 Interview with author, January 6, 2011.

27 Judith Layzer's case studies suggest that the expected benefits of EBM were oversold. Thus far, these policies have not achieved significant ecological restoration any more than policies formulated through conventional means. See Judith. A. Layzer, 2008, Natural Experiments, Cambridge, MA: MIT Press.

28 Christopher McGrory Klyza and David Sousa, 2008, American Environmental Policy, 1990–2006, Cambridge, MA: MIT Press.

29 USEPA, "Managing Urban Runoff," Office of Water, EPA841-F-96–004G.

30 Daniel Press, 2015, American Environmental Policy, Northampton, MA: Edward Elgar.

31 John A. Hoornbeek, 2011, Water Pollution Policies and the American States, Albany, NY: SUNY Press.

32 Seth Borenstein and Paul Rogers, 2000, "Clinton Picks Pollution Fight," San Jose Mercury News (July 12): 1A.

33 There were few examples of administrative action to control pollution during the Bush years. In one prominent example, the EPA approved new rules governing emissions from diesel engines.

34 See New York Times, 2008. "More Sadness for Appalachia," Editorial (October 20): A28. The Charleston Gazette also published a series of reports on this important decision and its consequences, http://wvgazette.com/static/series/mining.

35 At first, Salazar filed a lawsuit in U.S. District Court to block its implementation. When the suit was rejected, the new administration's only recourse was to begin a new rulemaking. Having committed publicly to reversing the rule, Salazar directed OSM to begin work on a new rule, which was proposed at the end of 2009. See Siobhan Hughes, "Obama Seeks to Reverse Mountaintop-Mining Rule," Wall Street Journal (April 28, 2009): A1.

36 The court found that the Bush-era rule had been issued without the necessary consultation with federal wildlife agencies, National Parks Conservation Association v. Jewell, D.D.C.1:09-cv-00115, February 20, 2014.

37 Name redacted. Congressional Research Service, 2017, "The Office of Surface Mining's Stream Protection Rule: An Overview" R44150. (January 11).

38 Full text of Trump's speech in Des Moines, Iowa, www.politico.com/story/2016/08/full-text-trump-227472.

39 Rapanos v. United States 547 U.S. 715 (2006).

40 Christy Leavitt, 2007, "Troubled Waters: An Analysis of Clean Water Act Compliance," Washington, DC: USPIRG Education Fund.

41 Charles Duhigg and Janet Roberts, "Rulings Restrict Clean Water Act, Foiling EPA," New York Times (February 28, 2010).

42 Quoted in Charles Duhigg and Janet Roberts, "Rulings Restrict Clean Water Act, Foiling EPA," New York Times (February 28, 2010).

43 Christy Leavitt, 2007, "Troubled Waters: An Analysis of Clean Water Act Compliance," Washington, DC: USPIRG Education Fund.

44 The bill was also vehemently denounced on television by Fox News Commentator Glenn Beck. See Charles Duhigg and Janet Roberts, "Rulings Restrict Clean Water Act, Foiling EPA," New York Times (February 28, 2010).

45 Eric Lipton and Coral Davenport, 2015, "Critics Hear E.P.A.'s Voice in 'Public Comments,'" New York Times (May 18).

46 Alan Wilson, 2015, "Attorney General Alan Wilson Joins Lawsuit Against EPA," www.scag.gov/archives/22110.

47 Coral Davenport, 2018, "Trump Plans to Begin E.P.A. Rollback with Order on Clean Water," New York Times (February 28).

48 Nationwide, only about 10 percent of CAFOs were subject to permitting because federal controls applied only to those facilities that intentionally discharged effluent into waterways. The requirement was waived for those that spilled waste because of stormwater. See Suzanne Herel, "EPA Squeaks Under Wire for Rules on Farm Runoff," San Francisco Chronicle (December 17, 2002): A1.

49 Office of Management and Budget, "Meeting Record Regarding EPA CAFO Rule," November 26, 2001, www.whitehouse.gov/omb/oira_2040_meetings_81.

50 Waterkeeper Alliance v. USEPA, 399 F. 3d 486 (2nd Cir. 2005).

51 Testimony of Kennedy, Robert F. Jr., before the Subcommittee for Administrative and Judicial Law of the Judiciary Committee, U.S. House of Representatives, February 4, 2009.

52 National Pork Producers Council v. U.S. EPA, 635 F. 3d 738 (5th Cir. 2011).

53 J. D. Alexander, quoted in Lora Berg, 2013, "EPA Chief Sends Regrets Yet Releases Sensitive Data Anyway," National Hog Farmer (February 22).

54 Another proposed amendment to the failed Senate Bill, sponsored by Kentucky Republican Rand Paul, would have prohibited the EPA from issuing any rules to clarify the jurisdictional scope of the Clean Water Act. "House, Senate Farm Bills Seek to Limit EPA Regulation of Agriculture," Inside EPA Daily Briefing (June 3, 2013), https://insideepa.com/.

55 Urbanized areas have a population density greater than 1,000 persons per square mile. The rule also allowed states to determine whether to require stormwater plans from non-urban municipalities with over 10,000 residents and a population density greater than 1,000 persons per square mile.

56 USEPA Office of Water, "Authorization Status for EPA's Stormwater Construction and Industrial Programs," <http://cfpub.epa.gov/npdes/stormwater/authorizationstatus.cfm.

57 Quoted in Pat Brennan, "New Sewer Rules Proposed by EPA Could Cramp City Finances," Orange County Register (January 11, 2001).

58 Ibid.

59 In a hearing on the proposal, Democratic Representative Bart Stupak of Michigan announced he had 77 co-sponsors for his proposed legislation to prohibit wastewater blending. Subcommittee on Water Resources and the Environment, House Committee on Transportation and Infrastructure, 109th Congress, April 13, 2005.

60 Testimony of Benjamin H. Grumbles, EPA Deputy Assistant Administrator for Water, before the Subcommittee on Water Resources and the Environment, U.S. House Committee on Transportation and Infrastructure, February 14, 2002, www.epa.gov/ocir/hearings/testimony/107_2001_2002/021402bg.PDF.

61 Benjamin H. Grumbles, EPA Deputy Assistant Administrator for Water, before the Subcommittee on Water Resources and the Environment, U.S. House Committee on Transportation and Infrastructure, February 14, 2002, www.epa.gov/ocir/hearings/testimony/107_2001_2002/021402bg.PDF.

62 Ronald Reagan, "State of the Union Address," January 25, 1984, www.presidency.ucsb.edu/ws/?pid=40205.

63 Alison Muscatine and Sandra Sugawara, "Chesapeake Bay Emerges as Reagan's Environmental Cause," Washington Post (October 7, 1984): E40.

64 Felicity Barringer, 2007, "A Plan to Curb Farm-to-Watershed Pollution of Chesapeake Bay," New York Times (April 13).

65 Judith. A. Layzer, 2012, The Environmental Case: Translating Values to Public Policy, 3rd edition, Washington, DC: Congressional Quarterly: 455.

66 Bill Clinton, "Remarks on the 25th Observance of Earth Day in Havre de Grace, Maryland," April 21, 1995, Public Papers of the President: William J. Clinton: 1995, Washington, DC, 1: 565.

67 David A. Fahrenthold, "Pollution Rising in Tributaries of Bay, Data Show," Washington Post (December 5, 2007): A1.

68 The May 1994 poll of 2,004 adults conducted by the University of Maryland Survey Research Center. See "Poll Shows People Disagree with Governments Over Who Pollutes the Bay," Washington Post (May 6, 1994): D3.

69 Howard R. Ernst, 2010, Fight for the Bay, Lanham, MD: Rowman and Littlefield.

70 David A. Fahrenthold, "Broken Promises on the Bay," Washington Post (December 27, 2008): A1.

71 George W. Bush, 2007, Weekly Radio Address, October 20.

72 Howard R. Ernst, 2010, Fight for the Bay, Lanham, MD: Rowman and Littlefield.

73 18,000 represents approximately 35 percent of all community water systems, serving approximately 77 million people. See M. B. Pell and Joshua Schneyer, 2016, "The Thousands of U.S. Locales where Lead Poisoning is Worse than Flint" (December 19), www.reuters.com/investigates/special-report/usa-lead-testing.

74 John H. Cushman, Jr, "Bill Would Give Water Customers Pollution Notice," New York Times (June 22, 1996): A1.

75 The AIDS community played a key role in forcing the issue onto the agenda after the 1993 outbreak. The San Francisco-based AIDS advocacy group ACT UP mobilized to call attention to the issue and pressure the EPA to implement new standards. This was unusual because the interest groups that normally involve themselves in drinking water standards tend to be environmental groups, such as the Natural Resources Defense

Council. Already politicized from the struggles during the 1980s over funding for research, the rights of AIDS patients and access to experimental drugs, AIDS activists saw microbial contamination of drinking water as a serious threat that disproportionately affected their community. The Federal Advisory Committee for the microbial rule included a representative from ACT UP.

76 Between the proposal in 1994 and 1998, when the final rule was issued, 30 toxicological studies on DBPs were published. Evidence from animal studies established the need to regulate chloroform, but also suggested that more information was needed to better understand the risk to humans posed by byproducts of chemical disinfectants in drinking water. Among the new evidence was a chloroform study disputing the linear effects model on which the standard was based.

77 The Chemistry Council challenged the decision not to rely upon the latest available science, and the NRDC intervened by filing a brief and making arguments of its own in support of the zero standard. In an unusual move, the EPA asked the three-judge panel to rule on the validity of the MCLG. See Frederick W. Pontius, 2000, "Chloroform: Science, Policy and Politics," Journal of the American Water Works Association 92(5): 12–18.

78 Chlorine Chemistry Council and Chemical Manufacturers Association v. EPA, U.S. District Court, Washington, DC 98–1627 (March 31, 2000).

79 Susan W. Putnam and Jonathan B. Weiner, 1995, "Seeking Safe Drinking Water" in Risk vs. Risk, edited by John D. Graham and Jonathan B. Weiner, Cambridge, MA: Harvard University Press.

80 Michael R. Edelstein and William J. Makofske, 1998, Radon's Deadly Daughters: Science, Environmental Policy, and the Politics of Risk, Lanham, MD: Rowman and Littlefield.

81 USEPA, 1987, "Unfinished Business."

82 A few site studies of ambient radon and in groundwater were in progress around the U.S. during the late 1970s and 1980s to determine the severity of radon concentrations. Considerably more research, including national studies, was not underway until the 1980s and 1990s. Hence, findings that were unavailable in 1991 altered regulators' perceptions of the problem after they were published. See National Research Council, 1999, Risk Assessment of Radon in Drinking Water, Washington, DC: National Academy Press.

83 Evidence from the more recent studies would have compelled the agency to devise a radon standard for drinking water, even without the legislative mandate. Because it has a cancer risk factor of one in 10,000, radon is one of the most carcinogenic substances regulated under the SDWA. See National Research Council, 1999, Risk Assessment of Radon in Drinking Water, Washington, DC: National Academy Press.

84 The American Water Works estimated the cost of implementing the 10 ppb standard at $1 billion annually. See Anita Huslin, "Debate Swells Over Arsenic in Water Supply," Washington Post (July 5, 2001): B1.

85 Timothy C. Benner, 2004, "Brief Survey of EPA Standard-Setting and Health Assessment," Environmental Science and Technology 38(13): 3457–3464.

86 Although regulated industries and local governments responded to OIRA's request for feedback, the majority of the 71 comments were provided by the Virginia-based Mercatus Center at George Mason University. According to the OIRA report, the authors were Mercatus Center Director and former OIRA Director Wendy Gramm, and Susan Dudley, a Mercatus research fellow who served as OIRA Director during Bush's second term. See Office of Information and Regulatory Affairs, 2001, "Making Sense of Regulation," Washington, DC: US Office of Management and Budget, www.whitehouse. gov/sites/default/files/omb/assets/omb/inforeg/costbenefitreport.pdf.

87 Richard N. L. Andrews, 2006, "Risk-Based Decision-Making: Policy, Science and Politics," in Environmental Policy: New Directions for the Twenty-First Century, 6th

edition, edited by Norman J. Vig and Michael E. Kraft, Washington, DC: Congressional Quarterly Press, 225.
88 Victoria Sutton, 2003, "The George W. Bush Administration and the Environment," Western New England Law Review 25(1): 221–42.
89 Brady Dennis, 2017, "Trump Administration Halts Obama-era Rule Aimed at Curbing Toxic Wastewater from Coal Plants," Washington Post (April 13).
90 Byron W. Daynes and Glen Sussman, 2010, White House Politics and the Environment, College Station, TX: Texas A&M Press.

Bibliography

Andrews, Richard N. L. 2006. "Risk-Based Decision-Making: Policy, Science and Politics." In *Environmental Policy: New Directions for the Twenty-First Century*, 6th edition, edited by Norman J. Vig and Michael E. Kraft. Washington, DC: Congressional Quarterly Press:225.

Barringer, Felicity. 2007. "A Plan to Curb Farm-to-Watershed Pollution of Chesapeake Bay." *New York Times* (April 13).

Benner, Timothy C. 2004. "Brief Survey of EPA Standard-Setting and Health Assessment." *Environmental Science and Technology* 38(13): 3457–3464.

Berg, Lora. 2013. "EPA Chief Sends Regrets Yet Releases Sensitive Data Anyway." *National Hog Farmer* (February 22).

Borck, Jonathan C., Cary Coglianese, and Jennifer Nash. 2008. "Evaluating the Social Effects of Environmental Leadership Programs." Working Paper of the Regulatory Policy Program, John F. Kennedy School of Government. Cambridge, MA: Harvard University.

Borenstein, Seth and Paul Rogers. 2000. "Clinton Picks Pollution Fight." *San Jose Mercury News* (July 12): 1A.

Brennan, Pat. "New Sewer Rules Proposed by EPA Could Cramp City Finances." *Orange County Register.* January 11, 2001.

Chlorine Chemistry Council and Chemical Manufacturers Association v. EPA. U.S. District Court, Washington, DC, 98–1627 (March 31, 2000).

Clinton, William J. "Executive Order 12866". Regulatory Planning and Review Online by Gerhard Peters and John T. Woolley, *The American Presidency Project.* www.presidency.ucsb.edu/node/227792, www.presidency.ucsb.edu/documents/executive-order-12866-regulatory-planning-and-review.

Clinton, William J. "Remarks on the 25th Observance of Earth Day in Havre de Grace, Maryland." April 21, 1995. *Public Papers of the President: William J. Clinton: 1995.* Washington, DC. 1: 565.

Clinton, William J. "The President's News Conference." April 18, 1995. *Public Papers of the President: William J. Clinton: 1995.* Washington, DC. 1: 547.

Congressional Research Service. 2017. "The Office of Surface Mining's Stream Protection Rule: An Overview" R44150 (January 11).

Cushman, John H., Jr. 1996. "Bill Would Give Water Customers Pollution Notice." *New York Times* (June 22): A1.

Davenport, Coral. 2018. "Trump Plans to Begin E.P.A. Rollback with Order on Clean Water." *New York Times* (February 28).

Daynes, Byron W. and Glen Sussman. 2010. *White House Politics and the Environment.* College Station, TX: Texas A&M Press.

Dennis, Brady. 2017. "Trump Administration Halts Obama-era Rule Aimed at Curbing Toxic Wastewater from Coal Plants." *Washington Post* (April 13).

DeWitt, John. 2004. "Civic Environmentalism," in *Environmental Governance Reconsidered*, eds Robert Durant, Rosemary O'Leary and Daniel Fiorino. Cambridge, MA: MIT Press.

Duhigg, Charles. "Clean Water Laws Neglected, at a Cost." *New York Times* (September 13, 2009): A1.

Duhigg, Charles and Janet Roberts. "Rulings Restrict Clean Water Act, Foiling EPA." *New York Times* (February 28, 2010).

Edelstein, Michael R. and William J. Makofske. 1998. *Radon's Deadly Daughters: Science, Environmental Policy, and the Politics of Risk*. Lanham, MD: Rowman and Littlefield.

Ernst, Howard R. 2010. *Fight for the Bay*. Lanham, MD: Rowman and Littlefield.

Fahrenthold, David A. "Broken Promises on the Bay." *Washington Post* (December 27, 2008): A1.

Fahrenthold, David A. "Pollution Rising in Tributaries of Bay, Data Show." *Washington Post* (December 5, 2007): A1.

Fiorino, Daniel J. 2006. *The New Environmental Regulation*. Cambridge, MA: MIT Press.

Freeman, A. Myrick, III. 2000. "Water Pollution Policy" in *Public Policies for Environmental Protection*, 2nd edition. Edited by Paul R. Portney and Robert N. Stavins. Washington, DC: 203.

Grumbles, Benjamin H. 2002. "Testimony before the Subcommittee on Water Resources and the Environment, U.S. House Committee on Transportation and Infrastructure" (February 14). www.epa.gov/ocir/hearings/testimony/107_2001_2002/021402bg.PDF.

Hanna-Attisha, Mona. 2018. *What the Eyes Don't See: A Story of Crisis, Resistance, and Hope in an American City*. New York: One World.

Harter, Philip J. 1982. "Negotiating Regulations: A Cure for Malaise." *Georgetown Law Journal* 71(1): 1–118.

Herel, Suzanne. "EPA Squeaks Under Wire for Rules on Farm Runoff." *San Francisco Chronicle*. December 17, 2002: A1.

Hoornbeek, John A. 2011. *Water Pollution Policies and the American States*. Albany, NY: SUNY Press.

Hughes, Siobhan. "Obama Seeks to Reverse Mountaintop-Mining Rule." *Wall Street Journal*. (April 28, 2009): A1.

Huslin, Anita. "Debate Swells Over Arsenic in Water Supply." *Washington Post* (July 5, 2001): B1.

Inside EPA Daily Briefing. "House, Senate Farm Bills Seek to Limit EPA Regulation of Agriculture" (June 3, 2013). https://insideepa.com.

Keiter, Robert B. 2003. *Keeping Faith with Nature*. New Haven: Yale University Press.

Kline, Benjamin. 2011. *First Along the River*. 4th edition. Lanham, MD: Rowman and Littlefield.

Klyza, Christopher McGrory and David Sousa. 2008. *American Environmental Policy, 1990–2006*. Cambridge, MA: MIT Press.

Kuehner, John C. "EPA-Fueled Upgrades for Sewers Could Cost Millions." *Cleveland Plain-Dealer*, January 16, 2001: 1B.

Layzer, Judith A.. 2008. *Natural Experiments*. Cambridge, MA: MIT Press.

Layzer, Judith A.. 2012. *The Environmental Case: Translating Values to Public Policy*, 3rd edition. Washington, DC: Congressional Quarterly, 455.

Leavitt, Christy. 2007. "Troubled Waters: An Analysis of Clean Water Act Compliance." Washington, DC: USPIRG Education Fund.

Lipton, Eric and Coral Davenport. 2015. "Critics Hear E.P.A.'s Voice in 'Public Comments'" *New York Times* (May 18).

Marcus, Alfred A., Donald A. Geffen and Ken Sexton. 2002. *Reinventing Environmental Regulation: Lessons from Project XL.* Washington, DC: Resources for the Future.

Milazzo, Paul C. 2006. *Unlikely Environmentalists: Congress and Clean Water, 1945–1972.* Lawrence, KS: University of Kansas Press.

Muscatine, Alison and Sandra Sugawara. "Chesapeake Bay Emerges as Reagan's Environmental Cause." *Washington Post* (October 7, 1984): E40.

National Performance Review. 1995. "Creating a Government that Works Better and Costs Less." U.S. Government Printing Office.

National Pork Producers Council v. U.S. EPA. 635 F. 3d 738 (5th Cir., 2011).

National Research Council. 1999. *Risk Assessment of Radon in Drinking Water.* Washington, DC: National Academy Press.

New York Times. 2008. "More Sadness for Appalachia" Editorial (October 20): A28.

Office of Information and Regulatory Affairs. 2001. "Making Sense of Regulation." Washington, DC: US Office of Management and Budget.www.whitehouse.gov/sites/default/files/omb/assets/omb/inforeg/costbenefitreport.pdf.

Office of Management and Budget. "Meeting Record Regarding EPA CAFO Rule." November 26, 2001. www.whitehouse.gov/omb/oira_2040_meetings_81>.

Pell, M. B. and Joshua Schneyer. 2016. "The Thousands of U.S. Locales where Lead Poisoning is Worse than Flint" (December 19). www.reuters.com/investigates/special-report/usa-lead-testing/.

Pontius, Frederick W. 2000. "Chloroform: Science, Policy and Politics." *Journal of the American Water Works Association* 92(5):12–18.

Press, Daniel. 2015. *American Environmental Policy.* Northampton, MA: Edward Elgar.

Putnam, Susan W. and Jonathan B. Weiner. 1995. "Seeking Safe Drinking Water" in *Risk vs. Risk,* edited by John D. Graham and Jonathan B. Weiner. Cambridge, MA: Harvard University Press.

Reagan, Ronald. "State of the Union Address." January 25, 1984. www.presidency.ucsb.edu/ws/?pid=40205.

Scheberle, Denise. 2004. "Devolution." In *Environmental Governance Reconsidered,* eds Robert Durant, Rosemary O'Leary and Daniel Fiorino. Cambridge, MA: MIT Press.

Sutton, Victoria. 2003. "The George W. Bush Administration and the Environment." *Western New England Law Review* 25(1):221–242.

USEPA. "Managing Urban Runoff." Office of Water. EPA841-F-96–004G.

USEPA. 2013. "National Rivers and Streams Assessment." EPA/841/D-13/001. February 28.

USEPA Office of Water. "Authorization Status for EPA's Stormwater Construction and Industrial Programs." http://cfpub.epa.gov/npdes/stormwater/authorizationstatus.cfm.

Vig, Norman J. 2006. "Presidential Leadership and the Environment." in *Environmental Policy: New Directions for the Twenty-First Century,* 6th edition, edited by Norman J. Vig and Michael E. Kraft. Washington, DC: Congressional Quarterly Press. 100–123.

Washington Post. "Poll Shows People Disagree with Governments Over Who Pollutes the Bay" (May 6, 1994): D3.

Waterkeeper Alliance v. USEPA. 399 F. 3d 486 (2d Cir., 2005).

Wilson, Alan. 2015. "Attorney General Alan Wilson Joins Lawsuit Against EPA." www.scag.gov/archives/22110.

Wines, Michael and John Schwartz. 2016. "Unsafe Lead Levels in Tap Water Not Limited to Flint." *New York Times* (February 8).

5

TOXIC COMMUNITIES

In the Trump Administration's first months, the EPA proposed a number of sweeping regulatory rollbacks, including mercury emissions standards for coal combustion. These actions, intended to revitalize the coal industry, would have the greatest impact on the predominantly low-income and minority communities downwind from coal-fired power plants. A 2012 National Association for the Advancement of Colored People (NAACP) report found the 6 million people living within three miles of coal-burning power stations nationwide had an average per capita income of $18,400. Looking at the 12 plants with the highest toxicity of emissions, the report determined that the 2 million people living within three miles of the facilities had a per capita annual income of $14,626, and three quarters of that population were people of color.[1] Therefore, any uptick in the use of coal would disproportionately impact minorities and the poor.

In the environmental decade of the 1970s, Congress enacted major statutes to control water and air pollution and protect endangered species. These were major concerns for a broad cross-section of the public, and they came to embody "the environment" to lawmakers. The less-visible problem of toxic waste would be tackled later in the decade. Absent from the agenda for at least another decade were questions about environmental risk and equity: how pollution affected specific communities and groups defined by race and class. Presidential leadership and activism by the executive branch were required before the problem of environmental justice was legitimated. In the face of legislative gridlock, Bill Clinton used a series of executive orders to advance an environmental equity agenda. He also took the initiative and introduced initiatives that combined public-private partnerships with traditional regulations to control toxic pollution. The Bush-Cheney Administration expanded upon these innovations, forging new partnerships and emphasizing voluntary standards while downplaying traditional regulations to such an extent that entire categories of toxic pollution were barely touched by federal regulation by the end of his presidency. Thus, the Obama Administration was mainly preoccupied with introducing traditional regulatory mechanisms to replace the voluntary initiatives of the Bush years. This chapter examines the effect of presidential administration on environmental equity. I distinguish between the problems of hazardous waste and pollution policies that subject entire communities to elevated risk. From hazardous waste siting and cleanup to air pollution "hot spots," presidential initiatives have shaped recent policies toward the distribution of environmental risk.

Environmental Justice

The concept of environmental justice has been expanded in recent years to encompass a variety of environmental risks, as they impact any size of geographically defined disadvantaged groups, from neighborhoods to entire nations.[2] Few thought about the issue before 1982, when residents of Warren County, North Carolina began to protest the decision by the state to site a landfill for the purpose of disposal of PCBs in their low-income, predominantly African-American county. In this era of a revived "New Federalism," it was becoming more common to delegate important decisions, such as citing hazardous waste facilities, to the states, as discussed further below. Civil rights and environmental groups found common ground in their mistrust of the Reagan Administration and shared suspicion that state governments would not enforce the national environmental laws.[3] The preponderance of locally unwanted land uses in impoverished communities is largely an accident of a federal system, in which states with larger industrial bases and a high standard of living also have stricter environmental controls, and are forced to seek host communities elsewhere.[4] The reasons for the proximity of these hazards to particular communities are complex, and researchers disagree on the role of race in the siting of toxic waste and other unwanted land uses.[5]

One of the groups that sided with the Warren Country protestors was the Commission for Racial Justice of the United Church of Christ. Concerns about broader implications of hazardous waste siting led the commission to release a report, *Toxic Waste and Race*. The 1987 report became the first attempt to address the question explicitly. In addition to its findings, the report concluded with a call for higher-quality and more reliable information about the risks to communities. These themes became rallying cries in 1991, when the First National People of Color Environmental Leadership Summit convened in Washington, DC. Delegates from communities of color from 50 states and U.S. territories drafted an "inclusive" environmental agenda.[6] The environmental justice movement is credited with democratizing environmentalism, broadening the community, and fighting for transparency and the disclosure of pollution data.

If it appeared to activists in the 1990s that environmental justice was an idea whose time had come, their hopes for legislation were dashed before long. In the 103rd Congress alone, six environmental justice bills were introduced, but none were enacted. Evan Ringquist argues that environmental justice lacked the necessary conditions for legislative activism. Agenda-setting is most likely to lead to policy change when there is a mobilized public, a favorable political climate, a powerful and supportive policy subsystem, and clearly defined causes and solutions.[7] As an issue on the formal agenda, environmental justice fell short on these criteria. Moreover, Congress had turned in a conservative direction, and lawmakers were not very receptive to claims of environmental racism or demands for new laws to promote equitable treatment. The Democrats had a broad environmental agenda following the 1992 election. As a candidate for president, Bill Clinton promised to expedite the mitigation of Superfund sites and to reform the program. Campaigning in Western Pennsylvania, Clinton addressed local concerns about hazardous emissions from a planned waste incinerator across the river in East Liverpool, Ohio, promising to review the project if elected. As president, he supported unsuccessful legislation to promote environmental justice, develop brownfields, and curb hazardous waste exports.

After those bills died in committee during his first year in office, Clinton moved ahead without new legislation and announced that a "top administrative priority" would be environmental equity.[8] This commitment was demonstrated in a pair of executive orders that expanded upon initiatives undertaken by his predecessor, George H. Bush. The first response to the EJ movement at the federal level was undertaken when the EPA established the Office of Environmental Justice in 1992. Its mission was to ensure that communities that were predominately low-income or minority receive equal treatment under the nation's environmental laws. In 1994, Clinton issued EO 12898, which created an interagency working group on environmental justice. Declaring that "all Americans have the right to be protected from pollution, not just those who can afford to live in the cleanest, safest communities," Clinton ordered agencies to evaluate their programs, policies, planning and participation procedures, as well as their enforcement and rulemaking activities. The order was grounded in legislative and administrative authority, specifically Title VI of the Civil Rights Act, which guarantees equal treatment under the law for minority populations, and bans discrimination against people regardless of race, color or national origin. The order also invoked a 1973 EPA rule, which extended Title VI to environmental programs, as well as a 1984 rule banning discriminatory actions by EPA employees. Taking the idea further, Clinton ordered federal agencies to integrate environmental justice into their missions. Building upon the foundation of NEPA, agencies were required "to identify and address the disproportionately high and adverse human health or environmental effects of their actions on minority and low-income populations" and to "develop a strategy for implementing environmental justice." The order affected thousands of industrial facilities as well as numerous state and local government initiatives, encouraging them to provide access to information and public participation by minority and low-income communities. Also significant, the order represented a first step toward collecting data to help track the transportation and distribution across state lines.[9]

Pursuant to EO 12898, the EPA released its 1995 Environmental Justice Strategy, which declared that no community or group should suffer disproportionately from adverse human health or environmental effects due to the agency's actions. In practice, the agency employed a dual strategy of promoting equity through civil rights legal actions in conjunction with discretionary authority under federal statutes.[10] The EPA began to pursue civil rights actions in 1993 with an investigation into the process of granting permits to polluting facilities in Louisiana's chemical alley. Subsequently, the EPA sued Borden Chemicals for improperly storing hazardous chemicals and polluting groundwater in nearby poor and minority communities. The number of civil rights complaints under Title VI increased during the 1990s, until Congress pushed back. In fiscal years 1999–2001, legislative riders prohibited the EPA from spending any money on environmental justice programs or pursuing lawsuits on the basis of the Civil Rights Act.[11]

The second strategy that remained available to the EPA is the vigorous use of administrative discretion under the nation's environmental laws. Most significant are the implementation and enforcement decisions that have the potential to mitigate or exacerbate the effects of past injustice. These include setting standards, creating opportunities for public participation, issuing permits to sources of hazardous emissions, and assessing penalties for violators. If the law is not applied in a fair and even-handed manner, those communities with a legacy of discrimination and economic hardship could be exposed to disproportionate environmental risk.

The culmination of EO 12898 was an Integrated Agency Action Agenda released in the final year of Clinton's presidency. The report gathered dust following the Bush transition, and there were few early signs of continuity. The newly-appointed EPA Administrator, Christine Todd Whitman, reaffirmed this inclusive aspect of the agency's mission in a 2001 memorandum declaring that "environmental justice is the goal to be achieved for all communities and persons," while omitting familiar references to ethnicity and income.[12] By the end of Bush's first term, the goal of integrating EJ into federal agencies was all but forgotten. On the tenth anniversary of Clinton's order, the EPA inspector general issued a report critical of the administration's inaction on EJ. Federal agencies had not followed through with their environmental justice missions, and the agency had not followed up in accordance with its directive to coordinate EJ strategies (EPA, 2004). The EPA was also criticized for watering down the definition of EJ by dropping references to minority and low-income populations, and the Government Accountability Office issued a scathing report to Congress critical of the EPA for neglecting environmental justice considerations in rulemaking.[13]

Sustainability and Equity

Whereas EO 12898 addressed environmental justice directly, Clinton used other presidential decrees to protect vulnerable populations from undue risk. Clinton issued EO 13045, Risks to Children, which established the Children's Public Health Advisory Committee in 1997 to assess environment and health risks to children and recommend policy changes.[14] Another executive order was given to implement commitments made by the Bush Administration at the Rio Summit in 1992. Pursuant to Agenda 21, Clinton issued EO 12852, which established the President's Council on Sustainable Development (PCSD) in 1993. The council was constituted from top officials from every federal agency, and for six years it deliberated on strategies to promote sustainability and released two reports. Unlike the accepted notion of sustainability dating back to the Stockholm Summit of 1972, the PCSD promoted a milder vision of "sustainable economic development" as its goal.

Before the council was disbanded in 1999, it recommended a new environmental management system that would be responsive to the concerns of the environmental justice movement, specifically, that "the distribution of pollution and environmental resources has been inequitable, to the disadvantage of low-income and minority communities."[15] The PCSD recommended that inequities in pollution could be prevented by creating mechanisms that "ensure that community representatives," other than elected officials, "are collaboratively involved in decisionmaking processes affecting the community." An optimal system, the commission suggested, should "identify risks and protect communities against disproportionate impacts." Based on the PCSD recommendations, Clinton issued EO 13148, outlining an Environmental Management System for the federal government to be in place by 2002. Clinton's sustainable development agenda was short-lived. Since it was based on a presidential directive rather than legislation, the government-wide sustainability mission did not live beyond Clinton's presidency, which was over after a few months. However, the initiative planted a seed that flourished. The federal government continued many of those initiatives, even though they were no longer called sustainability. Clinton's order was referenced in 2009, when new EPA Administrator Lisa Jackson issued a directive reiterating the agency's sustainability mission.

Repairing Superfund

The statutory framework that governs the transportation, use and disposal of hazardous substances originated in the latter half of the Environmental Decade of the 1970s. The Resource Conservation and Recovery Act of 1976 (RCRA) did not authorize the EPA to regulate unlicensed disposal sites that predated the law. Consequently, the legal framework provided no way to address the threats to soil and water quality from existing hazardous waste. Not long after Congress enacted RCRA, entirely new and unanticipated problems with abandoned waste sites began to dominate the environmental agenda. The revelation of toxic contamination and gross negligence in the Love Canal neighborhood near Niagara Falls, New York, raised questions about the environmental risk posed by the countless waste sites yet to be discovered. President Carter had called on Congress to close the gap in hazardous waste control by creating a new comprehensive liability law and a special fund to rehabilitate existing sites, but deep divisions over financing the program impeded the legislative debate. The Republican victories in the 1980 election provided the impetus for Congress to enact the law that created Superfund. Following Carter's defeat and the Democrats' loss of the Senate, the Comprehensive Environmental Response, Compensation, and Liability Act (CERCLA) of 1980, was passed in a lame-duck session. Assembled hastily with little time for debate, Superfund was seen as expensive and inefficient, since the statute required that its funds be spent on cleaning up hazardous waste sites and on legal fees while other environmental programs were underfunded. A number of studies provided support for the argument that the public health benefits of the program were not commensurate with its cost.[16]

When CERCLA was due for reauthorization in 1986, lawmakers took the opportunity to impose new mandates and place additional restrictions upon the EPA. It was one of several laws Congress rewrote to reduce the discretion they felt Reagan appointees had abused.[17] For example, when RCRA was reauthorized in 1984, Congress required the EPA to set standards for solid and liquid wastes in municipal landfills and impose new recycling requirements for the states. The Coastal Protection Act was amended in 1988 to include new permit requirements and penalties for ocean dumping. The Superfund Amendments and Reauthorization Act (SARA) not only increased funding for the program, it included strict requirements and timetables, even for routine implementation matters. Despite stricter language and deadlines, pitched battles over implementation ensued.[18] At the end of the 1980s, a RAND study found that at least 32 percent of Superfund's budget had been used for legal fees, and there was growing frustration about delays in site remediation.

Candidate Bill Clinton called Superfund "broken," and pledged to work with the 103rd Congress to reform the program. At the urging of the White House, business and environmental groups collaborated for months to develop a plan that all parties found satisfactory. Some in the administration, including Treasury Secretary Lloyd Bentsen, felt that retroactive liability should be repealed when the law was amended. EPA Administrator Browner maintained that the president would not support any bill that undermined the "polluter pays" principle, since that would shift cleanup bills onto the taxpayers.[19] The compromise bill was drafted and handed to Congress one year into Clinton's term. It included provisions for community participation in site selection, a brownfield redevelopment program, national standards to facilitate remediation, and a revolving fund financed by the private insurance industry, but retroactive liability was retained. In spite of broad support from the participating interest groups, the White

House-backed bill was rejected by Congress even though most of the difficult work had been done already. Doing away with retroactive liability was the main priority for Republicans. Anticipating gains in the 1994 midterm election, the party had less incentive to approve the White House-backed bill and more of a reason to leave Superfund reform to the next Congress.[20]

After reform legislation failed and Republicans won Congressional majorities in 1994, the Clinton Administration enacted many of the same reforms administratively. One change was to the liability formula the agency used. The other had to do with remedies. EPA began to balance costs and benefits of site remediation, allowing the use of the rehabilitated land to determine the appropriate level of "clean." The new Republican majority in the 104th Congress, which had emphasized regulatory reform in its "Contract with America," missed an opportunity to pre-empt Clinton's administrative reforms and create an improved Superfund in the process. There was enough common ground among critics of the program and supporters who wanted to see more cleanup activity at lower cost, with less money spent on litigation. However, the gridlock that blocked reform in the previous Congress persisted despite the change in leadership. The parties remained deadlocked over the cost, and no compromise was reached.[21] Clinton exploited Congressional inaction as he campaigned for re-election in 1996. On the eve of the Democratic convention in Chicago, Clinton visited Kalamazoo, Michigan, where he spoke about his unilateral leadership on changing the Superfund program. "I want an America in the year 2000 where no child should have to live near a toxic waste dump," Clinton declared, and called on Congress to approve nearly $2 billion for accelerated Superfund site cleanups and other hazardous waste programs.[22] In the fall presidential debates, he invoked the issue to provide a contrast with Congressional Republicans, including his opponent, former Senate Majority Leader Bob Dole:

> Their budget cut enforcement for the Environmental Protection Agency by a third. It cut funds to clean up toxic waste dumps with 10 million of our kids still living within four miles of a toxic waste dump, by a third. It ended the principle that the polluters should pay for those toxic waste dumps unless it was very recent.[23]

Clinton also used the debate stage to promise to redouble the effort, proposing tax credits to clean up two thirds of remaining toxic waste sites. Because of his advocacy, this was the only environmental issue mentioned in either one of the 1996 debates.

Even without reauthorizing legislation, the administrative reforms succeeded in revitalizing the Superfund program. Removing barriers and picking up the pace of site remediation were presidential priorities, and consequently, more sites were deleted from the National Priority List during his two terms than any comparable time period (see Figure 5.1). Nakamura and Church attribute the success of the reforms

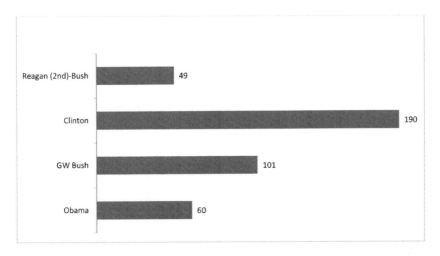

FIGURE 5.1 Superfund NPL Site Deletions, 1985–2016
Note: Very few sites were remediated before 1985 because the program began during Reagan's first term.

to a combination of leadership, effective management, and attention to incentives.[24] Not only had Clinton declared Superfund reform a goal for his administration, but so had several of his top-level appointees, including EPA Administrator Carol Browner and Justice Department officials in charge of environmental enforcement. Management was improved by providing clear instructions to regional offices, and finally, the administration altered the incentives by changing the performance measures for hazardous waste cleanup, rewarding officials for completing mitigation tasks.[25] By the end of Clinton's second term, the program's reputation had been turned around to such an extent that Browner's successor in the Bush Administration saw no reason to reverse the Clinton initiatives.[26]

Brownfields

One of the goals of the failed toxics reform legislation that Clinton pursued under his own authority was a program of waivers of Superfund liability to redevelop contaminated sites in urban areas. The brownfields strategy was intended to address a problem that was largely an unintended consequence of federal law, namely, the decline of aging industrial centers of Eastern and Midwestern cities where the worst industrial pollution was found. Many of these sites are unusable because of soil contaminated with neurotoxins, especially lead, chromium, mercury and benzene. By raising the cost of rehabilitating these abandoned sites, it reduced the cost of undeveloped land in rural and suburban areas by comparison. Therefore the laws that protected residents of inner cities from local toxic hazards tend to exacerbate their economic woes by discouraging new investment.

The Brownfields Action Agenda was announced in 1995. Acting under the authority of the Community Reinvestment Act of 1977, the administration encouraged cooperation between the EPA and lenders to ease financial liability and regulatory burdens for developers who sought CRA credit to rehabilitate urban industrial property.[27] There were pilot projects underway in 12 cities by the time Clinton began his uphill campaign for re-election in 1996. As described in Chapter 3, he turned Republican obstruction to his advantage as he launched a number of conservation initiatives administratively, helping to boost his environmental credentials while shaming a do-nothing Congress. Naturally, Clinton used some of this energy to claim credit for unilateral action on brownfields. Following the initial success of the EPA's program, Clinton called Congressional Republicans out for their criticism of Superfund. In March, Clinton visited Jersey, home to nationwide-high 107 Superfund sites, and urged Congress to approve $2 billion for tax credits to encourage brownfield redevelopment.[28] By the end of 1996, $200,000 grants had been awarded to support redevelopment projects in 40 cities.

The surplus of contaminated sites grew rapidly in the 1990s because of a large number of newly decommissioned military installations. In 1994, following three rounds of base closures, 95 former bases had been added to the NPL.[29] The Pentagon's Defense Environmental Restoration Program (DERP) was created under SARA to facilitate site remediation, but the program was badly underfunded. Clinton took action early in his first term to hasten the cleanup as well as the economic recovery of communities affected by base closures. He issued an executive order and established the Community Reinvestment Program to fast-track rehabilitation projects in 1993. Subsequently, he named OMB Director Alice Rivlin and CEQ Chair Kathleen McGinty to co-chair a task force, which recommended that redevelopment of bases be promoted by relaxing cleanup standards taking future uses into consideration.[30]

One consequence of the administration's regulatory activism on multiple fronts was the tension between certain initiatives. Clinton had an urban economic agenda that focused on relaxed environmental standards to promote development, sometimes undermining his own environmental justice goals.[31] Not surprisingly, the most vocal opposition to the brownfields initiative came from the environmental community. One of those voices was the National Wildlife Federation, which expressed skepticism about the possibility of more pollution under the guise of economic revitalization.[32] Another example of this tension is Clinton's advocacy for urban "empowerment zones." In this federal version of the Reagan-era enterprise zone idea that was adopted in 37 states, businesses were encouraged to bring jobs to blighted urban centers by relaxing taxes and regulations.

It was several years before Congress passed legislation that legitimized and expanded the Brownfields program that the Clinton Administration created under its own discretionary authority. Despite the program's popularity with lawmakers, efforts to pass a Brownfields bill stalled in the House as the issue became tied up with opposition to Superfund reauthorization.

Congress eventually enacted the Small Business Liability Relief and Brownfield Redevelopment Act of 2002. The law reduced liability for buyers of property and reduced liability for current property owners that did not contribute to contamination of the site, and provided exemptions for small businesses and residential property owners.[33] Although welcomed by the states, the Brownfield legislation was of limited value because funding for redevelopment projects was cut drastically from the federal budget.[34]

Fewer NPL sites were deleted during the Obama years, even though the program got an infusion in the stimulus package in 2009. In May 2017, incoming Administrator Scott Pruitt announced he would reprioritize and accelerate the remediation of the remaining 1,336 sites. He convened a 107-member task force, assembled from selected agency personnel and his political allies from his home state of Oklahoma. The task force held no public hearings, kept no records, and never disclosed the criteria used to select its members. Thirty days later, it produced a report listing 42 sites that should be prioritized because they could be developed for new construction, or because residents faced a threat from spreading pollution.[35] Those recommendations were immediately adopted by Pruitt.

Toxic Exports

Another unintended consequence of stricter disposal laws was a sharp increase in hazardous waste dumped overseas. By 1990, the U.S. was disposing of 70 million automobile batteries annually, and new restrictions forced disposal firms to locate new sites for roughly 70 million gallons of sulfuric acid waste and more than 1 billion pounds of lead.[36] Operators simply found it more cost-effective to ship used batteries to developing countries such as Mexico, Brazil and Taiwan, as opposed to the U.S. where the recycling practices were far safer and cleaner. The abandoned recycling facilities, while no longer emitting lead into their surrounding communities, remained as vacant Superfund sites. In 1990, a typical site of this kind would have been located in an urban area and, on average, would have cost $25 million to remediate.[37]

The elaborate regime of hazardous waste policies dating back to the 1970s offered no mechanism to restrict hazardous substances shipped overseas from the United States. After the lame duck Congress that passed CERCLA adjourned without addressing the problem, Carter decided to use an executive order to close this gap in the law. Five days before leaving the White House, Carter issued Executive Order 12264, which required the State Department to list hazardous substances restricted under U.S. law and to develop procedures to notifying foreign governments about possible hazards within contents of overseas shipments.[38] Hazardous exports would not be a priority for the Reagan Administration, and the U.S. sat on the sidelines in the 1980s as other developed nations ratified the Basel Convention. Carter's executive order was never implemented and it was reversed early in 1981.[39]

By the end of the 1980s, environmental justice emerged as a global issue. There was growing awareness that anyone could be affected by cases of neglect far away. Just two years after Bhopal, a toxic waste spill in the Rhine poisoned water in communities in nations downstream. There was also growing backlash to the practice of exporting hazardous waste to poor and politically powerless places around the globe. In one example that captured headlines in the U.S., the cargo ship *Khian Sea* roamed the oceans for two years in search of a place for its cargo of 14,000 tons of incinerator ash from Philadelphia after it was turned away from a New Jersey port. The toxic Flying Dutchman unloaded 4,000 tons in Haiti before it was ordered to leave. The remaining 10,000 tons of ash, containing high levels of lead, cadmium and other heavy metals, was believed to be dumped in the Indian Ocean sometime in 1988. In response to this and other incidents, the Basel Convention was ratified in 1989. The treaty became legally binding in 1992 without support from the U.S., which had previously committed to reduce hazardous waste exports under the 1976 Toxic Substances Control Act (TSCA).

One problem that symbolizes environmental justice on a global scale is the recycling of marine vessels. The domestic ship recycling industry moved overseas in the 1980s after safety, health and environmental regulations drove up the cost of dismantling ships in dry dock. Older military vessels containing asbestos, as well as high concentrations of heavy metals and PCBs, were increasingly sent to Bangladesh, India and Pakistan to be scrapped. This turned out to be illegal, as TSCA prohibits exports of products with concentrations of PCBs that exceeded 50 parts per million, but this prohibition went unenforced until 1994.[40] It would be another 15 years before Congress would enact legislation to require government-owned ships to be recycled domestically, but in a presidential directive, Clinton declared a moratorium on recycling government-owned vessels abroad. Even with the prohibition in force, privately-owned ships continued to wind up in overseas shipbreaking yards after they were sold to third parties. In 2012, the Indian Supreme Court sided with local officials and environmentalists who blocked the scrapyard-bound *Oriental Nicety*, an oil tanker formerly known as the *Exxon Valdez,* because of the toxic materials it contained.[41]

At the same time the U.S. wrestled with its exports of hazardous waste, there was new attention to toxic chemicals produced in the U.S. for use in overseas markets, especially pesticides. In 1989 Representative Mike Synar used Congressional hearings to call attention to the dumping of pesticides prohibited for domestic use in overseas markets. Despite this renewed attention to the problem, Congress was unable to pass legislation to control pesticide exports. Clinton worked with Congress to update domestic pesticide regulations with the Food Quality Protection Act of 1996, but Congressional approval of international treaties on toxics proved elusive. What Clinton did accomplish on waste management and pesticide controls took the form of international executive agreements, which were not subject to Senate ratification.

"Reinventing Government" Reconsidered

The Bush Administration preferred deregulatory and nonregulatory approaches. This meant relaxing regulations on some forms of hazardous waste, using voluntary agreements to address previously uncontrolled problems such as mercury emissions and coal ash, and creating new flexibility for regulated industries. These practices were employed from the dawning days of his presidency until his final month in office, when the EPA finalized rules allowing the burning of toxic waste for fuel. Reinventing government was all but discontinued after 2000, but parts of Clinton's reinvention program thrived under his successor. The aspects of reinvention that the Bush Administration embraced were the idea of partnerships and voluntary standards. Whereas the Clinton Administration used these instruments to complement traditional regulations, Bush saw them as more of a substitute for regulation.[42]

One early initiative was intended to create flexibility in the hazardous waste disposal requirements of RCRA. The law, which authorizes the EPA to list forms of waste as "potentially hazardous" to human health and the environment, has been faulted for raising the cost of treating and disposing of waste and discouraging recycling. In 2003, the EPA Office of Solid Waste launched the Resource Conservation Challenge to emphasize product stewardship, reduce the production of priority chemicals and cut the size of the overall waste-stream. EPA regional offices worked with local governments and businesses to establish voluntary programs with a goal of boosting recycling by 35 percent by 2008.[43]

Disclosure Policies and the Right to Know

There has been some success with a nonregulatory approach to environmental risk at the community level. One example is the Toxic Release Inventory (TRI). By publicizing instances of toxic pollution, the TRI has the potential to accomplish more than a traditional standards and enforcement regime. Making the data available online in the 1990s made the information more accessible, turning the TRI into one of the most cost-effective national environmental programs.[44] The legislation that created the TRI was established in response to demands from the community right-to-know movement in the 1980s. TRI was constructed upon a foundation of political activism and incremental legislation two decades in the making.

The Community Right-to-Know movement hit its stride at a time of transition for the broader environmental community. During the Reagan years, the largest national environmental organizations established a permanent beltway presence to defend the movement's policy gains from the previous decade. Even as the mainstream groups' influence began to wane, there was new energy from groups like Lois Gibbs' Citizens Clearinghouse for Hazardous Waste (CCHW).

Despite their loss of access in the early 1980s, activists scored victories at the local level. They succeeded in passing right-to-know laws in several cities, beginning with Philadelphia. Williams and Matheny attribute the movement's influence to three strategies. First, the movement facilitated community access to information about local hazards. Second, it linked grassroots groups around the country. Third, its primary location at the local level of the federal system gave the movement clout that it would not have enjoyed if it concentrated its lobbying in Washington.[45]

The first time a federal statute established the principle of Community Right-to-Know was when Superfund was reauthorized in 1986. The Emergency Planning and Community Right to Know Act (EPCRA) was passed into law as Title III of SARA. EPCRA encouraged state and local governments to develop plans for hazardous chemical releases and to provide local officials with records of hazardous chemicals present in their communities. The law followed a string of successes on the state level, and a major focusing event that caused federal lawmakers to take notice. The December 1984, disaster at the Union Carbide plant in Bhopal killed 5,000 and tens of thousands more were blinded in an accidental release of Methyl Isocyanate. Just six months later, there was a nonlethal leak at Union Carbide's twin plant near Institute, West Virginia. Reacting to fears of a large-scale disaster in the U.S., Congress included the disclosure provision when Superfund was reauthorized in 1986. Even though SARA had bipartisan support, there was more enthusiasm for EPCRA, which was approved by unanimous vote in the House of Representatives. It was a largely symbolic response to worst-case fears in the wake of a horrific focusing event.

Information disclosure requirements rest on the assumption that sources are more likely to police themselves if they know they are being monitored by the public. Information disclosure has thus been held up as an example of a non-regulatory alternative because communities are empowered to respond to specific threats and equipped to monitor polluters and pressure local officials.[46] When toxic releases occur, they are expected to report on the risks to the community or be labeled a poor citizen and face possible political and market reprisals. Furthermore, the TRI is credited with promoting environmental equity by providing community-level information to better facilitate collective action.[47] Despite the enhanced availability of toxic release data, opinions on the program's effectiveness are mixed. A comprehensive study of the TRI found that scarcely any citizen groups have used the data.[48] Less certain is the likelihood that the TRI will help disadvantaged communities where other political resources, such as money, influence and leisure time are in short supply. As Ackerman and Heinzerling put it, "'Not-in-my-backyard' protests against local hazards may work well for the wealthy, but not so well when the backyards belong to the poor or powerless."[49]

EPCRA had little immediate impact. The law required facilities to begin reporting disposal and releases of toxic chemicals into air, water and soil in 1987. The law specified 650 substances that would be subject to reporting requirements.

Firms meeting the eligibility requirements had to file an annual form "R," which was the first step toward building a national data archive. The TRI might not have achieved its full potential were it not for a pair of executive orders by President Clinton. In 1993, Clinton issued EO 12856, which furthered its implementation by making the federal government subject to EPCRA reporting requirements. In 1995, Clinton issued Executive Order 12969, "Toxic chemical release and community right-to-know," in 1995. Thanks to the widespread integration of information technology, the database became easily accessible, leading to the Toxic Release Inventory as we know it today.

Additional actions during the Clinton years gave the TRI enhanced relevance. In 1999, the EPA finalized a rule lowering the reporting threshold for persistent bioaccumulative toxins. The TRI soon became another political football. Whereas the Clinton Administration used executive orders and rulemaking to expand the scope and requirements of the program, the Bush Administration turned to the same tactics to scale it back. In 2006, the EPA issued a rule relaxing the reporting requirements. Without delisting any substances, the rule increased the reporting threshold for most of the 650 toxic substances from 500 pounds to 5,000 pounds. The rule drastically reduced the number of events reported, and dozens of toxic substances were no longer reported because their releases rarely exceeded 5,000 pounds.

Senator Barack Obama opposed the 2007 TRI rule and pledged to reverse it once he was elected president. Doing so would have required a lengthy new rulemaking process, diverting time and energy from Obama's other priorities at the start of his presidency. Since the Bush rule was unpopular with many in Congress, it was overturned as soon as the Democrats controlled both branches of government in 2009. During Obama's first frenetic weeks in office, Congress passed the Omnibus Appropriations Act of 2009, which contained language restoring the pre-2006 reporting requirements. Thus, the changes to the TRI were reversed faster than the administration could have done on its own, and the effect was more permanent.

Information Disclosure and Fracking

Information disclosure became the regulatory option of last resort for the approximately 13,000 new natural gas wells going online annually when Obama took office. The natural gas boom was driven by the availability of new technology to extract natural gas from the shale through hydraulic fracturing (fracking) and the absence of regulatory barriers at the federal level. As Chapter 3 discussed, accessing the untapped reserves beneath the shale was a key goal of the Bush/Cheney energy strategy. It was further encouraged after the election of Obama, who promoted natural gas production as a low-emission alternative to coal combustion. This part of Obama's energy policy was at odds with some of his environmental goals because the impact of the fracking method was not fully understood.

By 2009, grassroots activists were beginning to raise doubts about the safety of fracking in communities that experienced a rapid proliferation of natural gas wells, particularly in the Northeast, Texas, and the Western Plains. Fracking was blamed for a series of small earthquakes because seismic activity was considered unlikely in these regions. Of greater concern was the risk to groundwater that fracking posed. Actors within the Obama Administration gingerly took actions to bring fracking under federal regulation. In 2011, the EPA began a comprehensive study of the impact of natural gas drilling. The study identified several stages in the hydraulic fracturing process where drinking water is at risk: the withdrawal of surface and groundwater to create fracking fluids, the potential for accidental chemical spills, well injection, wastewater flowback, and the treatment and disposal of wastewater.[50] The data collection may be of little consequence because Congress exempted fracking from the disclosure rules of the Safe Drinking Water Act in 2005 amid industry assurances that the practice was safe.

The administration had other regulatory tools at its disposal besides information-gathering. With SDWA unavailable, opponents turned their focus to the toxic chemicals used in fracking. Soon after the drinking water study was announced, the EPA responded to an Earthjustice petition to develop reporting rules under TSCA, proposing requirements for industry disclosure of accidental releases and health effects of fracking chemicals. The major law that regulates the manufacture and distribution of chemicals, TSCA authorizes the EPA to maintain a public database of existing chemicals and to identify substances that present an "unreasonable" health risk. The law requires chemical producers to submit information about their products, including records about health effects. The EPA must be notified about new substances ninety days before they go onto the market to allow time for screening, and the EPA could suspend production or sale of a new substance if notification requirements are not met. The EPA is allowed to impose labeling requirements, and even chemicals if necessary. The proposed disclosure requirements for the fracking industry were far short of the strict regulations that environmentalists preferred, but industry groups howled in protest nonetheless. The American Chemistry Council complained the TSCA is too blunt an instrument to regulate fracking, arguing that a program of voluntary reporting would be more effective because it would apply to the operators.[51]

Coal Combustion Residuals

A nonregulatory approach was initially attempted to control coal combustion residuals (CCRs). Commonly known as coal ash, this was the largest hazardous waste stream not subject to RCRA when the Bush/Cheney Administration formed a partnership with the coal industry. Coal ash was not a national priority in the 1970s, when policymakers designed an elaborate regulatory framework to protect public health from sulfur dioxide and reduce the haze and acid precipitation downwind from coal plants. Americans were accustomed to thinking of air pollution from coal combustion as a national problem, even though burning coal also generates solid waste containing concentrated levels of arsenic, lead and heavy metals. Most coal ash is deposited in landfills, where it poses risks to communities adjacent to the disposal sites. For example, a coal ash landfill in Northwest Indiana was cited for spoiling groundwater in the town of The Pines, and in 2012, the EPA also agreed to investigate claims that the landfill was the source of the town's elevated radiation levels.[52]

Since ash stored on-site was not subject to federal regulation, many power generation sources in poor and rural areas pile their waste in uncovered retention ponds. There are more than 1,000 of these sites, mostly in Appalachia and the Midwest, although a few were located near rural coal-fired power plants in Western states. The burden of improper disposal practices is placed almost exclusively upon the communities adjacent to the impoundments. Periodically, Appalachian watersheds are fouled when these impoundments overflow after heavy rains. In the worst cases, the dams give way, unleashing catastrophic floods like the 1972 Buffalo Creek disaster that killed 125 people and wiped out an entire town. In dry regions, wind-scattered ash poses additional health risks. One example can be found 350 miles outside of Las Vegas, where Nevada's Moapa Band of Paiutes Reservation is afflicted with extraordinarily high rates of asthma and heart disease due to coal ash dust from the nearby Reid-Gardner Generating Station.

The Bush Administration's Coal Combustion Products Partnership (C2P2) sought to reduce the coal ash waste stream at a time when no national standards for disposal were in place. Most other activities of the coal industry were under some form of federal regulation by the 1970s, and there was a general expectation that the public would be protected from the health and safety effects of coal combustion as well as coal mining. The coal partnership promoted beneficial use of two types of CCRs (bottom ash and fly ash) by recycling them as concrete and gravel fill. As with the Resource Conservation Challenge, the emphasis was on voluntary compliance with industry-standard disposal practices, but it included no enforceable standards or sanctions.

The prospects for a nonregulatory approach to coal ash diminished in 2008, with the election of President Obama followed by the largest coal slurry spill in U.S. history. An unlined retention pond at a Tennessee Valley Authority power plant collapsed three days before Christmas, flooding 300 acres and two rivers near Kingston, Tennessee, with nearly 1 billion gallons of CCRs. In the estimated $1.2 billion cleanup, most of the toxic slurry was disposed in a landfill 350 miles to the south in Uniontown, Alabama. Robert Bullard noted that the disposal site was located in Alabama's poorest county, which also happened to be 88 percent African-American. In the wake of the Kingston disaster, Lisa Jackson, President Obama's newly appointed EPA administrator, ordered a safety review of coal ash impoundments and announced new regulations by the end of 2009. When the report was released several weeks later, Jackson announced that at least

44 of the nation's highest-risk coal ash impoundments had integrity problems and reiterated her commitment to propose new regulations. Since an act of Congress would be required to classify coal ash as hazardous waste, the EPA ultimately proposed two alternatives to regulate coal ash using its administrative authority. The first alternative was to classify coal ash as a "special" waste. Then, under Schedule C of RCRA, the EPA could phase out the impoundments and require new disposal practices in compliance with RCRA. The second alternative was to regulate coal ash under Schedule D as non-hazardous waste. Under this approach, there would be national standards, but the states would regulate disposal. Since coal ash would be treated as solid waste, the practice of disposal in conventional lined landfills would continue.

By the end of 2012, the EPA received more than 400,000 comments, mostly in support of the stricter alternative. Industry preferred the second alternative, which offered little change from the status quo. In a dynamic reminiscent of the late 1990s, industry allies in Congress mobilized to pre-empt the EPA, and the administration was on the defensive. In 2011, the new Republican majority in the House of Representatives passed a bill giving states the authority to regulate coal ash instead of the EPA, but the bill went no further. A final Coal Combustion Residuals rule was issued in 2015, establishing minimum federal standards for the disposal of coal ash and required companies to monitor groundwater and make their data publicly available. In 2018, the Trump Administration proposed a revision to the standards, suspending groundwater monitoring requirements if it is determined there is no potential for contamination of aquifers.

Airborne Toxics

Another consequence of coal combustion not controlled by the original Clean Air Act is mercury pollution. It remained unregulated a decade after the Clean Air Act Amendments of 1990 required the EPA to set standards for 189 hazardous air pollutants, including mercury. The Clinton Administration, which had long emphasized market-based regulations, briefly considered a cap-and-trade market as an alternative to command-and-control standards for power plants, which were responsible for at least 40 percent of mercury emissions. EPA Administrator Carol Browner thought such a market would be illegal because mercury is a human neurotoxin, therefore an emissions trading scheme threatened to poison communities downwind of the source. White House Chief of Staff John Podesta agreed with Browner, and the policy was not pursued. In December 2000, as George W. Bush prepared to occupy the White House, the Clinton team punted. The EPA issued a regulatory finding that the Clean Air Act's language on hazardous air pollutants required mercury emissions to be regulated at the source. Even though the Clinton Administration had not fulfilled the law's ten-year old mandate by promulgating regulations, it took steps to prevent the kind of market-based program that industry wanted in the future.

The midnight ruling on mercury seemed to constrain the incoming Bush Administration, which sought to relax regulations on the coal industry wherever possible. Since Section of 112 of the 1990 Clean Air Act required standards based on best available technology (BAT) for hazardous air pollutants, anything less stringent than a technology-based standard was precluded by the signature legislative achievement of Bush's father. Consequently, the Bush Administration proposed legislation in 2002, known as "Clear Skies," to control mercury, sulfur and nitrogen oxide pollution from power generation sources. The plan was to set emission standards and create a system of marketable permits to achieve the reductions. The administration's proposal was controversial because allowing sources to increase their mercury output by accumulating credits would create the potential for "hot spots." Since mercury emissions tend to accumulate within 100 miles of their source, the adjacent communities would be at increased risk. Moreover, the proposal would have allowed sources to "bank" their emissions by borrowing from future allotments to increase short-term output. Senators from both parties insisted that limits on carbon dioxide be included in any legislation. Once a bipartisan group of senators lined up behind an alternative bill that not only set mercury emissions standards but prohibited both trading and banking, support for the administration bill withered, and neither one was enacted.

The failure of Clear Skies prompted the administration to try to sidestep the legislative process. Key provisions were lifted from the bill and proposed as regulatory changes. According to OIRA Director John Graham, when Congress failed to enact the administration's bill, Bush authorized the EPA to create a cap-and-trade market for sulfur dioxide and nitrogen oxide under its own authority.[53] The Clean Air Interstate Rule, finalized in 2005, established a trading market for states east of the Mississippi River. Although the program was expected to reduce sulfur dioxide emissions by approximately 70 percent by 2015, some environmentalists were concerned that it would allow high-emission plants to avoid new pollution controls, as well as increases in pollution that would be concentrated downwind. Another objection was that the program allowed sources to bank sulfur dioxide credits, which had the potential to undermine the existing acid rain program. These objections were noted when the rule was remanded in a unanimous decision of the U.S. Circuit Court in D.C. in 2008. The Obama Administration proposed a modified version of the rule, imposing stricter emission standards and restrictions on banking, but even this was vacated by the U.S. Circuit Court in D.C. in 2012, and the EPA was told to start over.[54]

Mercury is a neurotoxin released into the air by coal combustion. The EPA estimates that annually, more than 75,000 newborns have a higher risk of learning disabilities due to in-utero mercury exposure. Because of the Clean Air Act's requirements for hazardous air pollutants, a cap-and-trade market for mercury could not be imposed through the administrative process. Hence, the EPA proposed a rule downgrading mercury from a hazardous substance. The redefinition was controversial within the EPA, because Deputy Administrator Jeffrey

Holmstead, a political appointee, circumvented the science advisory committee. This cleared the way for the 2005 Clean Air Mercury Rule (CAMR) to create an emissions trading market.[55] Several state governments and environmentalists sued over the EPA over the delisting of mercury from Section 112, and also because the proposed rule would have resulted in less mercury reduction than the law required for hazardous pollutants.

The CAMR case illustrates another key aspect of Bush's administrative strategy: centralization of federal policy through preemption. As with several other Bush administration regulations on food safety and consumer product safety, the mercury rule included language that precluded states from setting their own mercury standards for coal combustion.[56] Once the D.C. Circuit Court vacated the rule in 2008, the states that challenged it proceeded to adopt their own mercury standards nonetheless. The EPA appealed the ruling, but the petition was dropped after Obama's election. Obama's first EPA Administrator, Lisa Jackson, stated her intent to propose a new technology-based standard, promising "...a toxics regulation, so it probably means setting limits."[57] The EPA finally issued its final mercury and air toxics standards (MATS) in 2011, more than 20 years after the Clean Air Act's mandate, although waivers were offered for sources that needed additional time to comply.

Changing the Math, Ignoring the Science

Repealing MATS was one of the specific "action items" coal baron Robert Murray had prescribed for the Trump Administration in early 2017. Rescinding or rewriting the standards would have been problematic because they had taken effect already. The Trump administration used a novel approach: Instead of amending the substance of the rule, it changed the assumptions behind the cost-benefit analysis to justify its implementation. The EPA announced it would weigh only the *direct* benefits of mercury reduction against the cost of implementing the rule.[58] *Co-benefits* of MATS—the clean air benefits of retiring coal-fired power plants and the health benefits from a decline in other pollutants such as particulate matter—would no longer be counted. Consequently, the benefits of cutting mercury pollution would range between $4 million and $6 million annually, which is significantly less than the $7.4 billion to $9.6 billion it costs to implement the rule. The Obama Administration's EPA had counted an additional $80 billion in co-benefits from MATS. The rollback was proposed in December 2018 by Andrew Wheeler, Trump's second EPA administrator and a former coal lobbyist. By then, the power industry had largely met the MATS restrictions either by retiring old coal-fired power plants or installing new technology to capture mercury and other pollutants. However, this changes the calculation behind MATS implementation and could extend the lives of older coal plants still in operation.

Trump appointees had more retrenchment strategies beyond reversing individual regulations. One was to undermine the science at the foundation of existing and future regulations. The EPA proposed a rule in 2018, "Strengthening Transparency in Regulatory Science," placing restrictions on the type and amount of studies that could be used to set standards for toxic and hazardous substances. The proposal instructed regulators to consider only those studies that use publicly available data, to promote transparency and replicability.[59]

This would preclude most epidemiological studies and public health studies which use confidential information on human subjects, including the Harvard "Six Cities" study, which provided evidence of the mortality risk from particulate matter in 1993.[60] The proposal also raised concerns about how radiation would be regulated. One passage instructed regulators to "consider threshold models across the exposure range," which raised concerns among environment and public health professionals that the EPA might be looking to establish a "safe" level of radiation for the workplace.[61] This would be a significant change, because the EPA has operated under the theory that no level of radiation is safe. Establishing a threshold would mean regulations to protect workers from low-level radiation may no longer be cost-effective

Enforcement actions declined under Trump, but this trend was underway before his arrival. Reflecting declining budgets, the number of civil suits filed by the EPA declined during Obama's second term. Nowhere was this more visible than in the EPA's pattern of criminal enforcement. As Figure 5.2 shows, there was a gradual decline in the number of criminal investigations initiated by the EPA. The steepest drop occurred in 2017, after the agency was staffed with Trump Administration appointees with ties to the chemical industry. One such appointee was David Dunlap, who was previously employed by Koch Industries as an expert on chemical regulations. Dunlap was hired to work for the EPA's Office of Research and Development, and because the office of director was left vacant, Dunlap as deputy director was free to set priorities for the office without

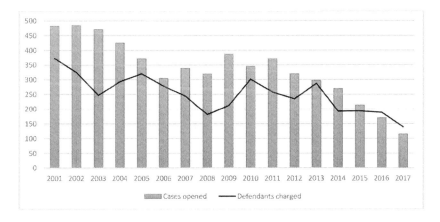

FIGURE 5.2 EPA Enforcement Actions, 2001–2017

being subject to Senate confirmation. The drop in enforcement is also apparent from the fines collected; in the first two years of the Trump Administration, fines were down 85 percent from the equivalent period of Obama's presidency.

One ongoing problem with toxics policy was the statute itself. TSCA was well-intentioned but it lacked teeth, and it was plagued with loopholes which allowed dangerous substances to remain on the market even when the risks were known. Strengthening TSCA was a priority for Senator Frank Lautenberg for more than a decade, but it was not until well after Lautenberg's death in 2013 that Congress managed to pass a TSCA reform bill. The Frank Lautenberg Chemical Safety for the 21st Century Act was a rare bipartisan accomplishment, finally giving the EPA authority and resources to advance chemical safety. The law set requirements and deadlines for the EPA to test 20 substances at a time, and it specified faster action for three toxic chemicals: methylene chloride and NMP, which are ingredients in paint strippers, and TCE, a solvent used in dry cleaning.

The victory was short-lived because it was easily undermined by Trump appointees. In 2017, Pruitt hired Nancy Beck to be Deputy Administrator for Chemical Safety, and Liz Bowman to be Associate Administrator for Public Affairs. Both were previously employed by the American Chemistry Council.[62] The EPA had recommended bans on the three toxic chemicals prioritized by Congress, but in 2017, citing costs to industry, Pruitt announced the bans were indefinitely postponed. He also issued a directive that narrowed TSCA's scope. Even though safety assessments were mandated for these and other toxic chemicals, the EPA was directed to focus only on the health effects of direct exposure.[63] The effects of air, water and ground contamination would be ignored. As with the MATS for coal combustion, the administration was technically in compliance with the law, but the scope of the assessment was restricted to produce industry-friendly outcomes.

Conclusion

Beginning in the 1990s, the president and bureaucracy took the initiative and adopted new regulatory approaches to control toxic and hazardous substances in the face of legislative gridlock. In the absence of comprehensive reform of Superfund, RCRA and TSCA, the executive branch was empowered to pursue its own initiatives, subject to frequent reversals with changes of administration. Clinton regulatory activism advanced an environmental justice agenda through executive orders. He also accomplished with rulemaking what he failed to achieve through the legislative process: reform and revitalization of the troubled Superfund program.

This gridlock has prevented Congress from closing gaps in the nation's toxics laws. After Congress failed to enact toxic waste legislation in the 1990s, the Clinton Administration's administrative reform initiative provided a path forward.

Due to Reinventing Government and Project XL, new partnerships were also formed, leading to successful state-level recycling and brownfield reclamation programs. Even though Clinton's major environmental policy legacy, administrative reform, was partially dismantled by his successor, one aspect of his reform agenda thrived. Bush embraced the idea of voluntary standards, creating new public-private partnerships to control hazardous waste such as coal ash, emphasizing them to such an extent that they became a substitute for regulation.

Both the Bush and Obama Administrations sought to federalize toxics policy, albeit for different reasons. Even though George W. Bush had the benefit of a Republican Congress for most of his presidency, he avoided new environmental legislation, preferring to use administrative actions to pursue his agenda. When it appeared that Congress would not act on the administration's "Clear Skies" proposal, Bush pursued his agenda through administrative channels, instructing the EPA to design cap-and-trade markets for sulfur and nitrogen oxides, as well as mercury, which required a significant rule change because of the Clean Air Act's restrictions on hazardous air pollutants. As with many other Bush Administration regulatory proposals, the mercury rule contained language to preempt the states, many of which had already enacted their own pollution standards. These initiatives illustrate the limitations of the administrative strategy; neither one survived the ensuing legal challenges.

Among the host of environmental challenges that confronted Obama were forms of hazardous pollution that had been neglected by his predecessors. The Bush Administration's nonregulatory approach to coal ash had proven inadequate, and the cap-and-trade proposal for mercury could not withstand legal challenge. Furthermore, Bush Administration's tacit support for hydraulic fracturing by the natural gas industry aroused the concerns of grassroots activists around the country, thus creating pressure on the new president to deal with fracking as a federal issue. This placed the Obama Administration in the position of responding to new demands from the environmental community to deal with emerging risks from hazardous pollution. Further constraining the administration was a series of Congressional acts that limited EPA authority. Due to these riders, the EPA was prohibited from pursuing civil rights actions, setting standards for coal ash or protecting drinking water from fracking. Without legislative authority to deal with these new problems directly, the administration was forced to use its authority creatively under existing statutes.

The Trump Administration pursued an agenda of policy retrenchment that favored the chemical and fossil fuel industries. Rather than simply delay, suspend and change certain regulations, the Trump Administration worked within the rules accomplished much of what it wanted by working within the existing framework. Ignoring the science and changing the math, the Trump appointees would effectively rig their assessments to produce the conclusions they sought. By focusing on the costs and ignoring most of the benefits, they concluded regulations on mercury emissions and toxic chemicals used in consumer products and were too costly to justify. They would also use this strategy with the global greenhouse gas regulations inherited from the Obama Administration, and the consequences would be profound.

Notes

1 NAACP, 2012, "Coal Blooded,"www.naacp.org/wp-content/uploads/2016/04/Coa lBlooded.pdf.
2 David Schlosberg, 2013, "Theorising Environmental Justice: The Expanding Sphere of a Discourse," *Environmental Politics* 22 (1): 37–55.
3 Charles Lee, 1993, "Beyond Toxic Wastes and Race" in *Confronting Environmental Racism: Voices from the Grassroots*, edited by Robert D. Bullard, Boston: South End Press.
4 Barry Rabe, 1994. *Beyond NIMBY*. Washington, DC: Brookings.
5 See Evan J. Ringquist, 2004, "Environmental Justice" in *Environmental Governance Reconsidered*, edited by Robert F. Durant, Daniel J. Fiorino and Rosemary O'Leary, Cambridge, MA: MIT Press.
6 Charles Lee, 1993, "Beyond Toxic Wastes and Race" in Confronting Environmental Racism: Voices from the Grassroots, edited by Robert D. Bullard, Boston: South End Press.
7 The conditions for agenda-setting and policy change were identified by John Kingdon in Agendas, Alternatives and Public Policies, 2nd edition. See Charles Tiefer, 1994, *The semi-sovereign presidency: the Bush administration's strategy for governing without Congress*, Boulder: Westview Press.
8 Evan J. Ringquist, 2004, "Environmental Justice" in *Environmental Governance Reconsidered*, edited by Robert F. Durant, Daniel J. Fiorino and Rosemary O'Leary, Cambridge, MA: MIT Press.
9 Barry Rabe, 1994, *Beyond NIMBY*, Washington, DC: Brookings.
10 Evan J. Ringquist, 2004, "Environmental Justice" in *Environmental Governance Reconsidered*, edited by Robert F. Durant, Daniel J. Fiorino and Rosemary O'Leary, Cambridge, MA: MIT Press.
11 Ibid.
12 USEPA, 2001, "Administrator Whitman Reaffirms Commitment to Environmental Justice" (August 21), https://archive.epa.gov/epapages/newsroom_archive/newsrelea ses/41a2df9798d627a185256aaf0067e435.html.
13 U.S. Government Accountability Office, 2005, "EPA Should Devote More Attention to Environmental Justice When Developing Clean Air Rules," GAO-05–289 (July), www.gao.gov/new.items/d05289.pdf.
14 The committee published reports in 2000 and 2003, and after a long hiatus, 2011.
15 Marc A. Eisner, 2007, *Governing the Environment*. Boulder, CO: Lynne Rienner.
16 See James T. Hamilton and W. Kip Viscusi, 1999, *Calculating Risks: The Spatial and Political Dimensions of Hazardous Waste Policy*, Cambridge, MA: MIT Press.
17 Reagan's first EPA Administrator, Anne Gorsuch had resigned after a brief and stormy tenure following a Congressional inquiry over a mishandled hazardous waste cleanup in Southern California in 1982. See Norman J. Vig, 2006, "Presidential Leadership and the Environment" in *Environmental Policy: New Directions for the Twenty-First Century*, 6th edition, edited by Norman J. Vig and Michael E. Kraft, Washington, DC: Congressional Quarterly Press. 100–123.
18 For example, SARA authorized the EPA to design a contingency plan for site cleanups, and to force the matter, the amended law set an April 1988 deadline for the final rule. The EPA not only missed the deadline, but only released the proposed rule after eight months of delay in which the Reagan OMB pressured the EPA to use a narrower threshold for cancer risk instead of the standard risk level used by the agency for other regulations. See Conrad B. MacKerron, "Has OMB Been Out of Bounds on EPA?," *Chemical Week* (March 22, 1989): 8.
19 Linda Feldmann, 1994, "Clinton Plans to Overhaul Superfund Law," *Christian Science Monitor* (January 24).

20 The inability of Congress to reach a compromise on a bill that makes even a modest improvement to the Superfund program has been explained as a form of "destructive competition." As Ray Squitieri argues, political considerations transformed the legislative debate from a positive-sum game into a zero-sum game because one side withholds support out of fear that the other side will take credit for the compromise. See Ray Squitieri, 2004, "Good and Bad News from Washington" in *Painting the White House Green*, edited by Randall Lutter and Jason F. Shogren, Washington, DC: Resources for the Future. 166–179.

21 Far from reforming the program, the one action taken concerning Superfund was a reduction in funding for cleanup in the 1995 Rescissions Act.

22 Martha T. Moore, "Clinton Proposes $1.9B for Toxic Waste Programs," *USA Today* (August 29, 1996): 4A.

23 William J. Clinton, Presidential Debate in Hartford Online by Gerhard Peters and John T. Woolley, The American Presidency Project, www.presidency.ucsb.edu/node/217284.

24 Robert T. Nakamura and Thomas W. Church, 2003, *Taming Regulation*, Washington, DC: Brookings.

25 Ibid.

26 Ibid.

27 The revised rules to allow the Brownfield waivers under the CRA were finalized May 4, 1995 (60 FR 22156).

28 Allison Mitchell, "Clinton Asks Tax Breaks for Toxic-Waste Cleanup," *New York Times* (March 12, 1996): A17.

29 Kenneth N. Hansen, 2004, *The Greening of Pentagon Brownfields*, Lanham, MD: Lexington.

30 Federal Facilities Policy Group, "Improving Federal Facilities Cleanup" (October 1995).

31 Norman Miller, 2009, *Environmental Policy*, 2nd edition, New York: Routledge.

32 Judith Evans, "Cleaning Up the Nation's Brownfields: Critics Want Assurances," *Washington Post* (November 25, 1995): E1.

33 Marc A. Eisner, 2007, *Governing the Environment*, Boulder, CO: Lynne Rienner.

34 Barry Rabe, 2007, "Environmental Policy and the Bush Era: The Collision Between the Administrative Presidency and State Experimentation," *Publius* 37(3): 413–431.

35 Michael Biesecker, "EPA Says Superfund Task Force Created by Pruitt Kept No Records of Meetings," *Chicago Tribune* (December 20, 2017).

36 Center for Investigative Reporting, 1990, *Global Dumping Ground*, Washington, DC: Seven Locks Press.

37 Ibid.

38 The order required the State Department to work with the Commerce Department to identify substances to be placed on a Commodity Control List. Anticipating industry objectives, it provided for the Commerce Department to "consult with affected industries" and "consider the likely effects of proposed controls on the export performance of the U.S. before imposing controls." It also contained language to "assure that trade secrets or other confidential commercial or financial information are not disclosed to foreign governments," unless authorized by law. See Lloyd Cutler, 1981, Memorandum to President Carter, "Proposed Executive Order Re Export of Hazardous Substances," January 14 (JCL: Staff Offices-Cutler Box 74).

39 Reagan's CEQ Chair. A. Alan Hill stated that the policy "would have imposed export controls through a cumbersome regulatory program costly to both the public and private sector." See Allan A. Hill, 1982, Letter to Natural Resources Defense Council, November 23 (JCL: Natural Resources-CEQ Box 1).

40 Basel Action Network, 2012, "Industrial Capabilities of North America: A Report on 'Green' Ship Recycling in the United States, Canada and Mexico" (November), www.ban.org.

41 Philip Siddarth, "Exxon Valdez Barred Entry into India Shipbreaking Yard," *Businessweek* May 10, 2012.
42 Marc A. Eisner, 2007, *Governing the Environment*, Boulder, CO: Lynne Rienner.
43 USEPA, 2003, Position paper "Beyond RCRA."
44 See Michael E. Kraft, Mark Stephan and Troy Abel, 2011, *Coming Clean*, Cambridge: MIT Press.
45 Bruce A. Williams and Albert Matheny, 1995, *Democracy, Dialogue and Environmental Disputes*, New Haven, CT: Yale University Press.
46 Richard H. Thaler and Cass R. Sunstein, 2008, 2009, *Nudge: Improving Decisions about Health, Wealth and Happiness*, New York: Penguin.
47 Michael E. Kraft, Mark Stephan and Troy Abel, 2011, *Coming Clean*, Cambridge: MIT Press.
48 Ibid.
49 Frank Ackerman and Lisa Heinzerling, 2004, *Priceless: On Knowing the Price of Everything and the Value of Nothing*, New York: The New Press.
50 USEPA, 2012, "Study of the Potential Impacts of Hydraulic Fracturing on Drinking Water Resources: Progress Report" (December), www.epa.gov/hfstudy/pdfs/hf-report20121214.pdf.
51 Bridget DiCosmo, "Industry Warns Planned TSCA Rule Inappropriate for Fracking Chemicals," Inside EPA (April 17, 2012), www.insideepa.com.
52 Lauri Keagle, 2013, "EPA: Study of Residential Yards in The Pines Planned," *Northwest Times* (January 3), www.nwitimes.com/news/local/porter/duneland/the-pines/epa-study-of-residential-yards-in-the-pines-planned/article_33096fc2-4980-5242-bb2c-3a2b641ebff4.html.
53 John Graham, 2010, *Bush on the Homefront: Domestic Policy Triumphs and Setbacks*, Bloomington: Indiana University Press.
54 *North Carolina v. EPA*, 531 F. 3d 896 (D.C. Cir. 2008).
55 Clean Air Mercury Rule, 70 Fed. Reg. 28,606 (May 18, 2005) 40 C.F.R. pts. 60, 72, 75.
56 It was not the first time a president who championed the cause of states' rights considered pre-empting a state's hazardous substance regulations. The Reagan Administration was in a similar position in 1986 when California voters approved Proposition 65, an initiative that imposed strict notification requirements for hazardous substances. Industry fears of a ripple effect, such as other states following California's lead, or Congress responding with a tougher federal Food and Drug Act, created pressure for the Reagan Administration to pre-empt the law. A task force of the Domestic Policy Council was formed to consider a request by the Processed Food Association that the FDA issue regulations to pre-empt California's law and fend off "...a crazy-quilt of regulations." See Domestic Policy Council, "Memorandum on Federal Preemption of State Food and Drug Regulations," July 27 1988.
57 Inside EPA, 2009, "Mercury Rule Reversal" (February 6), www.insideepa.com.
58 Lisa Friedman, Kendra Pierre-Louis and Livia Albeck-Ripka, 2018, "Law That Saved the Bald Eagle Could be Vastly Reworked," *New York Times* (July 19).
59 Proposed Rule, "Strengthening Transparency in Regulatory Science," F.R. 83 18768–18774 (April 30, 2018).
60 Amanda Paulson, 2018, "Are Proposed EPA Rules a Move Toward Transparency or an Attack on Science?," *Christian Science Monitor* (April 25).
61 Proposed Rule, "Strengthening Transparency in Regulatory Science," F.R. 83 18768–18774 (April 30, 2018).
62 Beck was hired to lead the EPA's Chemical Safety Office after the withdrawal of Trump's original nominee, Michael Dourson, a lobbyist for the Halogenated Solvents Industry Alliance. See Sheila Kaplan, 2017, "EPA Delays Ban on Uses of Hazardous Chemicals," *New York Times* (December 19).
63 Eric Lipton and Hiroko Tabuchi, 2018, "For Lease: Land to Drill on, Dirt Cheap," *New York Times* (November 28).

Bibliography

Ackerman, Frank and Lisa Heinzerling. 2004. *Priceless: On Knowing the Price of Everything and the Value of Nothing*. New York: The New Press.

Basel Action Network. 2012. "Industrial Capabilities of North America: A Report on 'Green' Ship Recycling in the United States, Canada and Mexico" (November). www.ban.org.

Biesecker, Michael. "EPA Says Superfund Task Force Created by Pruitt Kept No Records of Meetings." *Chicago Tribune* (December 20, 2017).

Center for Investigative Reporting. 1990. *Global Dumping Ground*. Washington, DC: Seven Locks Press.

Clean Air Mercury Rule. 70 Fed. Reg. 28,606 (May 18, 2005). 40 C.F.R. pts. 60, 72, 75.

Clinton, William J. Presidential Debate in Hartford. Online by Gerhard Peters and John T. Woolley, *The American Presidency Project*. www.presidency.ucsb.edu/node/217284.

Cutler, Lloyd. 1981. Memorandum to President Carter, "Proposed Executive Order Re Export of Hazardous Substances." January 14 (JCL: Staff Offices-Cutler Box 74).

DiCosmo, Bridget. "Industry Warns Planned TSCA Rule Inappropriate for Fracking Chemicals." Inside EPA (April 17, 2012). www.insideepa.com.

Domestic Policy Council. "Memorandum on Federal Preemption of State Food and Drug Regulations." (July 27, 1988).

Eisner, Marc A. 2007. *Governing the Environment*. Boulder, CO: Lynne Rienner.

Evans, Judith. "Cleaning Up the Nation's Brownfields: Critics Want Assurances." *Washington Post* (November 25, 1995): E1.

Federal Facilities Policy Group. "Improving Federal Facilities Cleanup" (October 1995).

Feldmann, Linda. 1994. "Clinton Plans to Overhaul Superfund Law." *Christian Science Monitor* (January 24).

Friedman, Lisa, Kendra Pierre-Louis, and Livia Albeck-Ripka. 2018. "Law That Saved the Bald Eagle Could be Vastly Reworked." *New York Times* (July 19).

Graham, John. 2010. *Bush on the Homefront: Domestic Policy Triumphs and Setbacks*. Bloomington: Indiana University Press.

Hamilton, James T. and W. Kip Viscusi. 1999. *Calculating Risks: The Spatial and Political Dimensions of Hazardous Waste Policy*. Cambridge, MA: MIT Press.

Hansen, Kenneth N. 2004. *The Greening of Pentagon Brownfields*. Lanham, MD: Lexington.

Hill, A. Allan. 1982. Letter to Natural Resources Defense Council. November 23 (JCL: Natural Resources-CEQ Box 1).

Inside EPA. 2009. "Mercury Rule Reversal." (February 6). www.insideepa.com.

Kaplan, Sheila. 2017. "EPA Delays Ban on Uses of Hazardous Chemicals." *New York Times* (December 19).

Keagle, Lauri. 2013. "EPA: Study of Residential Yards in The Pines Planned." *Northwest Times* (January 3). www.nwitimes.com/news/local/porter/duneland/the-pines/epa-study-of-residential-yards-in-the-pines-planned/article_33096fc2-4980-5242-bb2c-3a2b641ebff4.html.

Kraft, Michael E., Mark Stephan and Troy Abel. 2011. *Coming Clean*. Cambridge: MIT Press.

Lee, Charles. 1993. "Beyond Toxic Wastes and Race" in *Confronting Environmental Racism: Voices from the Grassroots*, edited by Robert D. Bullard. Boston: South End Press.

Lipton, Eric and Hiroko Tabuchi. 2018. "For Lease: Land to Drill on, Dirt Cheap." *New York Times* (November. 28).

MacKerron, Conrad B. "Has OMB Been Out of Bounds on EPA?." *Chemical Week* (March 22, 1989): 8.

Miller, Norman. 2009. *Environmental Policy*, 2nd edition. New York: Routledge.

Mitchell, Allison. "Clinton Asks Tax Breaks for Toxic-Waste Cleanup." *New York Times* (March 12, 1996): A17.

Moore, Martha T. 1996. "Clinton Proposes $1.9B for Toxic Waste Programs." *USA Today (August 29): 4A.*

NAACP. 2012. "Coal Blooded." www.naacp.org/wp-content/uploads/2016/04/Coa lBlooded.pdf.

Nakamura, Robert T. and Thomas W. Church. 2003. *Taming Regulation.* Washington, DC: Brookings Institution Press.

North Carolina v. EPA. 531 F. 3d 896 (D.C. Cir. 2008).

Paulson, Amanda. 2018. "Are Proposed EPA Rules a Move Toward Transparency or an Attack on Science?." *Christian Science Monitor* (April 25).

Rabe, Barry. 1994. *Beyond NIMBY.* Washington, DC: Brookings.

Rabe, Barry. 2007. "Environmental Policy and the Bush Era: The Collision Between the Administrative Presidency and State Experimentation." *Publius* 37(3): 413–431.

Ringquist, Evan J. 2004. "Environmental Justice" in *Environmental Governance Reconsidered*, edited by Robert F. Durant, Daniel J. Fiorino and Rosemary O'Leary. Cambridge, MA: MIT Press.

Schlosberg, David. 2013. "Theorising Environmental Justice: The Expanding Sphere of a Discourse." *Environmental Politics* 22(1): 37–55.

Siddarth, Philip. "Exxon Valdez Barred Entry into India Shipbreaking Yard." *Businessweek* (May 10, 2012).

Squitieri, Ray. 2004. "Good and Bad News from Washington" in *Painting the White House Green*, edited by Randall Lutter and Jason F. Shogren. Washington, DC: Resources for the Future. 166–179.

Thaler, Richard H. and Cass R. Sunstein. 2008, 2009. *Nudge: Improving Decisions about Health, Wealth and Happiness.* New York: Penguin.

Tiefer, Charles. 1994. *The semi-sovereign presidency: the Bush administration's strategy for governing without Congress.* Boulder: Westview Press.

U.S. Government Accountability Office. 2005. "EPA Should Devote More Attention to Environmental Justice When Developing Clean Air Rules." GAO-05-289 (July). www. gao.gov/new.items/d05289.pdf >.

USEPA. 2001. "Administrator Whitman Reaffirms Commitment to Environmental Justice" (August 21). https://archive.epa.gov/epapages/newsroom_archive/newsreleases/ 41a2df9798d627a185256aaf0067e435.html.

USEPA. 2003. Position paper, "Beyond RCRA."

USEPA. 2012. "Study of the Potential Impacts of Hydraulic Fracturing on Drinking Water Resources: Progress Report" (December). www.epa.gov/hfstudy/pdfs/hf-rep ort20121214.pdf.

Vig, Norman J. 2006. "Presidential Leadership and the Environment" in *Environmental Policy: New Directions for the Twenty-First Century*, 6th edition, edited by Norman J. Vig and Michael E. Kraft. Washington, DC: Congressional Quarterly Press. 100–123.

Williams, Bruce A. and Albert Matheny. 1995. *Democracy, Dialogue and Environmental Disputes.* New Haven, CT: Yale University Press.

6

CLIMATE CHANGE

The day after Thanksgiving 2018, the federal government issued a report on the anticipated effects of climate change. It was a dire forecast for the 21st century, anticipating climate-related disasters causing hundreds of billions of dollars in damage.[1] Prepared by 300 scientists, working with 13 federal agencies and executive departments, the report provided the most detailed assessment of climate impacts upon the nation's economy, coastlines, infrastructure, public health. To no one's surprise, the White House downplayed the report. As a candidate, Donald Trump questioned the science of climate change, and the first two years of his presidency were devoted to aggressively rolling back Obama-era climate policies, and he continued to disregard and sometimes mock scientific reports that contradicted his own policies. The peculiar thing about this controversy was that the 2018 National Climate Assessment, was a joint publication of his own administration's science agencies and executive departments.

The 1,656-page report, which the White House hoped to discredit, was the latest quadrennial assessment on the effects of climate change mandated in 1990 by an act of Congress and President George H.W. Bush. The assessment was already underway when Trump took office, and its existence was likely a surprise to Trump's transition team, which had not bothered to attend briefings by departing Obama appointees. Rather than censor the report, the White House tried to ignore it. In 2005, the Bush White House ignited a controversy when a climate report was edited to minimize the human role in greenhouse gas (GHG) emissions, giving the story more publicity than it would have received otherwise. Trump took the opposite approach by preemptively releasing the report on the Friday after Thanksgiving Day; typically the slowest news day of the year.[2] This strategy backfired because the story about the report was picked up by news organizations around the world, shared widely on social media, and dominated the headlines. Thus, the administration was in the unusual position of trying to discredit its own report. A White House issued a statement said the assessment was "based on the most extreme scenario," and suggested the next quadrennial assessment would provide "balance." Some Trump advisers were more direct. A member of Trump's EPA transition team called the report "made-up hysteria," and the product of the "deep state," referring to bureaucrats and scientists inside federal agencies Trump believed were resisting his policies.[3]

There had been many examples of incoming presidents rejecting the initiatives of their predecessors. In the case of the Obama/Trump transition, the stakes were higher because there were now policies in place to reduce GHG emissions, and they were at risk of repeal by the new administration. As this chapter explains, the failure of Congress to enact climate legislation increased the pressure on the Obama Administration to pursue a solution through the regulatory process. Consequently, the most significant climate policies of the Obama years took the form of administrative rules and agreements, notably the Clean Power Plan and the Paris Agreement, which were vulnerable to reversal by the Trump administration.

The climate was a low priority for policymakers even in the bygone era of Congressional dominance, when major legislative accomplishments targeted pollution and wilderness preservation. As evidence of global warming mounted, the problem remained largely unknown and poorly understood outside the scientific community. In 1965, President Lyndon B. Johnson delivered an alarming message to Congress: "This generation has altered the composition of the atmosphere on a global scale through radioactive materials and a steady increase in carbon dioxide from the burning of fossil fuels."[4] However, the global warming issue lacked the salience of the first-generation problems of pollution and land conservation that inspired Americans to take to the streets on Earth Day. The climate remained on the back burner in part because the issue is highly technical, and the effects seem remote by contrast with the powerful symbolism of clean air and clean water.

The problem began to permeate the public consciousness in the 1980s, then the hottest decade on record. News reports raised awareness of the greenhouse effect and its connection with fossil fuel combustion and disasters such as droughts and wildfires. Concerned members of Congress invited climate experts to hearings at the Capitol, where they were told of the potential catastrophe that could result if the world's industrial nations failed to reduce their greenhouse gas emissions and if deforestation continued. Congressional leaders were instrumental in organizing the first World Conference on the Changing Atmosphere in 1988, which produced a warning that industrialized nations must reduce their output of CO2 by 20 percent by 2005 in order to stabilize the earth's temperature.[5] Encouraged by the successful creation of a regime to protect stratospheric ozone, the United Nations Environment Program convened the Intergovernmental Panel on Climate Change (IPCC) in order to review the growing scientific literature and make its own recommendations. Global warming became an issue in the 1988 campaign, as Vice President George H.W. Bush took up the issue as part of his effort to distance himself from the Reagan Administration's environmental record. Bush pledged to be "the environmental president" and to meet the "Greenhouse Effect" with the "White House Effect." The climate was squarely on the national agenda by the end of the decade, driven in part by the IPCC report of 1990 and popular works like *The End of Nature*. [6] However, there were a number of impediments to achieving comprehensive climate legislation. As this chapter shows, leadership from the White House was responsible for a variety of moderate policy initiatives during the following two decades.

Climate Politics in the 1990s

There were significant factors that prevented the bold public action needed to address the problem. Unlike the first-generation environmental problems of pollution and natural resource conservation, climate change policies require broad international commitments. Even though the U.S. was by far the largest source of global greenhouse gas (GHG) emissions, there was little incentive for Congress to support unilateral reductions, and no national consensus to do so.

Just a few years earlier, Congress and the public embraced a different global problem that required international commitments. In order to save the ozone layer, lawmakers responded with binding legislation in the 1990 Clean Air Act to enforce targets established by the Montreal Protocol. However, the problem of climate change was not as simple as the depletion of stratospheric ozone, nor was the solution as simple. As Shellenberger and Nordhaus observe, there is a tendency for environmental advocates to look to that regime as a model for climate change policy without appreciating the differences between the problems. The authors are equally skeptical of efforts to address the problem with a cap-and-trade regime resembling the sulfur dioxide market of the 1990 Clean Air Act because the problem is so much more complex and global in scope than acid rain.

Because climate politics emerged after the business countermobilization of the 1980s, there has been balancing of interests preventing any meaningful legislative response. Beginning with George H.W. Bush's commitment to the Kyoto framework at the 1992 Rio Summit, policy responses to climate change have more often originated within the executive branch. Beginning in the 1990s, most significant policy actions were undertaken administratively with the authority granted by existing statutes.

Global warming was only one of several environmental policy issues on Bill Clinton's crowded platform when he was first elected in 1992. Clinton approached the problem much as he did mining and grazing reform, asking for legislation to alter the incentives for overproduction and waste. His proposal for a BTU (British Thermal Unit) Tax in the 103rd Congress was intended as a conservation and deficit-reduction policy, although it would have had the indirect effect of reducing carbon emissions. After passing narrowly in the House, the bill perished because of the opposition of senators from energy-producing states.

As Congress grew even less hospitable, Clinton had some success as he turned instead to his administrative strategy. In addition to key appointments he made to environmental posts (see Chapter 2) Clinton created new positions that helped coordinate policy initiatives and further his climate agenda both at home and abroad. For example, noted environmentalists were placed within the State Department. Former Colorado Senator Tim Wirth was named Undersecretary of State for Global Affairs and Eileen Claussen was named Undersecretary for Oceans and International and Scientific Affairs. In addition, he created the position of Federal Environmental Executive. Initially, this position was tasked with helping the General Services Administration procure recycled and sustainable products, but the office took on a more significant role later on.

Clinton signed the Kyoto Protocol in 1997, pledging a 7 percent reduction below 1990 levels by 2012, but never submitted the treaty to the Senate out of concern that it could not receive 67 votes for ratification. Rather, the Senate unanimously passed a resolution, cosponsored by Democrat Robert Byrd and Republican Chuck Hagel, demanding that the U.S. not ratify the treaty unless developing nations agreed to emissions limits as well. Rather than allow the treaty to go down in defeat, Clinton bypassed the senate by reaffirming the U.S. commitment in the form of an executive agreement.

Regarding domestic policy, Clinton's approach to climate change was essentially nonregulatory, emphasizing new public-private partnerships. Methane emissions from the livestock and mining industries were brought under control through AgStar and the Coalbed Methane Outreach Program, and EnergyStar created marketplace incentives for energy-efficient appliances. Shortly before the Kyoto summit, Clinton gave an address on climate change in Washington DC, where he praised voluntary commitments by corporations, for instance British Petroleum's promise to reduce GHG emissions to 10 percent below 1990 levels by 2000.[7] Eventually, there would be more stringent regulatory action to promote energy efficiency, but this came late in Clinton's presidency when the political risks were lowest. One month before the 2000 election, the Department of Energy proposed efficiency standards for four types of household appliances: washing machines, water heaters, air conditioners and heat pumps.

One other significant development with implications for GHG regulation was the EPA announcement in 1998 that it would begin to enforce the Clean Air Act's New Source Review (NSR) requirements. In 1999, the Justice Department and Attorneys General from Northeastern states filed lawsuits against nine utility companies for failing to make required improvements when making major repairs that extended the operation of their facilities. These suits languished in limbo after Clinton left office and the Bush Administration announced plans to modify the NSR requirements. The delay had the effect of allowing older, highly polluting power plants to continue operating at high capacity for several more years without resolution.

Changing Politics and Policy Continuity in the Bush Years

The climate was not a major issue in the 2000 election. However, that campaign was notable for the fact that for the first time, both candidates talked about climate change when it was discussed in one of the presidential debates. The issue was acknowledged by one or both candidates in presidential debates from 2000 to 2008. Table 6.1 lists the mentions of global warming and climate change by candidates in those debates. In 2000, both George W. Bush and his Democratic opponent, Vice President Al Gore, discussed the need to reduce CO2 emissions. The only way Bush distinguished himself from Gore was in qualifying his position to suggest that the Kyoto framework was not the best strategy for achieving those reductions.

Since Bush and Gore took similar positions on greenhouse gas regulation, this had the likely effect of neutralizing any advantage enjoyed by Gore, whose environmental credentials were already well known, in a very close race. The issue remained prominent enough on the campaign agenda of the two subsequent election cycles. However, the point of disagreement between the candidates was President Bush's handling of GHG reductions. In 2004, Democratic challenger John Kerry criticized Bush for ignoring climate science. Most mentions of climate change in 2008 were made by Republican John McCain, who sought to distance himself from Bush and pledged to address GHG emissions if elected. Climate change was a less salient issue in presidential politics after the Bush years. It was never mentioned in the three debates between Obama and Republican Mitt Romney in 2012. Table 6.2 lists the mentions of climate change in the 2016 presidential debates. It was only raised by Democrat Hilary Clinton, whereas Republican Donald Trump mocked the idea the idea that climate change was a national security risk.

Shortly after assuming office, Bush reversed his CO2 pledge and announced the U.S. would withdraw from the Kyoto Protocol. Whereas U.S. policy toward multilateralism took a U-turn, domestic policy remained largely unchanged. Since the Clinton Administration's approach to GHG reduction was predominantly one of partnerships and voluntary agreements, there were no major regulatory policies to roll back. After promising to reduce carbon emissions but rejecting Kyoto, Bush reverted to the voluntary approach that was used as a substitution for regulation of other pollutants (see Chapter 5). In the place of the emission standards that would have been required under Kyoto, Bush proposed the voluntary Climate Leaders program in 2002. The partnership was based on voluntary commitments by corporations to make an inventory of greenhouse gasses produced in their electrical generation and pledge to make improvements. It was controversial because of its intended goal of reducing "carbon intensity," or tons of CO2 for the amount of economic activity, as opposed to actual emissions.[8]

The administration's 180-degree shift on GHG emissions angered the environmental community and even some moderate Republicans, including his own EPA Administrator. Early in 2001, Christine Todd Whitman attended a summit in Europe,

TABLE 6.1 Mentions of Climate and Global Warming in Presidential Debates, 2000–2008

10/11/2000	
Gore	There was a study just a few weeks ago suggesting that in summertime the north polar ice cap will be completely gone in 50 years. Already many people see the strange weather conditions that the old timers say they've never seen before in their lifetimes.
Gore	I'm really strongly committed to clean water and clean air, and cleaning up the new kinds of challenges like **global warming**.
Bush	I tell you one thing I'm not going to do is I'm not going to let the United States carry the burden for cleaning up the world's air. Like **Kyoto Treaty** would have done.
Gore	I disagree that we don't know the cause of **global warming**. I think that we do. It's pollution, carbon dioxide, and other chemicals that are even more potent, but in smaller quantities, that cause this.
Bush	...some of the scientists, I believe, Mr. Vice President, haven't they been changing their opinion a little bit on **global warming**?
Bush	Look, **global warming** needs to be taken very seriously, and I take it seriously. But science, there's a lot – there's differing opinions. And before we react, I think it's best to have the full accounting, full understanding of what's taking place.
9/30/2004	
Kerry	You don't help yourself with other nations when you turn away from the **global warming** treaty, for instance...
Kerry	...he's not acknowledging the truth of the science of stem-cell research or of **global warming** and other issues.
Kerry	They pulled out of the **global warming**, declared it dead, didn't even accept the science. I'm going to be a president who believes in science.
Bush	Had we joined the **Kyoto treaty**, which I guess he's referring to, it would have cost America a lot of jobs. It's one of these deals where, in order to be popular in the halls of Europe, you sign a treaty. But I thought it would cost a lot...
9/26/2008	
McCain	Nuclear power is not only important as far as eliminating our dependence on foreign oil but it's also responsibility as far as **climate change** is concerned.
McCain	I have opposed the president on spending, on **climate change**, on torture...
Obama	As it is our national security and the issue of **climate change** that's so important.
10/7/2008	
Obama	We're not going to be able to deal with the climate crisis if our only solution is to use more fossil fuels that create **global warming**.
McCain	Whether it being a variety of other issues, working with Senator Lieberman on trying to address **climate change.**
McCain	I have disagreed strongly with the Bush administration on this issue. I traveled all over the world looking at the effects of **greenhouse gas emissions**, Joe Lieberman and I. And I introduced the first legislation, and we forced votes on it.
McCain	I have disagreed with the Bush administration. I have disagreed with leaders of my own party. I've got the scars to prove it. Whether it be bringing **climate change** to the floor of the Senate for the first time...

TABLE 6.2 Mentions of Climate and Global Warming in Presidential Debates, 2016

9-26-2016	
Clinton:	"Some country is going to be the clean- energy superpower of the 21st century. Donald thinks that **climate change** is a hoax perpetrated by the Chinese. I think it's real."
Trump:	"The single greatest problem the world has is nuclear armament, nuclear weapons, not **global warming**, like you think and your—your president thinks."
10-9-2016	
Clinton:	"So I have a comprehensive energy policy, but it really does include fighting **climate change**, because I think that is a serious problem.
10-19-2016	
Clinton:	"I want us to have the biggest jobs program since World War II, jobs in infrastructure and advanced manufacturing. I think we can compete with high-wage countries, and I believe we should. New jobs and clean energy, not only to fight **climate change**, which is a serious problem, but to create new opportunities and new businesses."

where she reassured G-8 ministers of Bush's commitment to capping carbon emissions. Only upon to her return to Washington did she learn that he had decided to withdraw from the Kyoto Protocol.[9] The closest Congress came to a solution was the Climate Stewardship Act of 2003, co-sponsored by Republican Senator John McCain and Democratic Senator Joe Lieberman. However, there was little support in the Republican Congress for the bipartisan bill to reduce GHG emissions by creating a market for tradable permits, which had been the preferred approach to implement the Kyoto goals. In his second term, Bush addressed emissions indirectly by signing the Energy Act of 2007, which encouraged conservation with new efficiency mandates. However, his executive actions would prove far more consequential than legislation.

In other respects, the Bush Administration followed an extension of the Reagan formula for restraining the bureaucracy. The White House used an administrative strategy to contain new GHG regulations while continuing the non-regulatory approach of the Clinton years. Political appointees at the EPA, CEQ, and OMB curtailed and delayed the development of GHG regulations that were deemed harmful to fossil fuel-derived energy production and key industries. These controls by the White House were so intrusive that agency scientists were silenced and government reports on the effects of climate change were censored. In 2005, a whistleblower revealed that CEQ Chief of Staff Philip Cooney had edited the text of a final report by the U.S. Government's Climate Change Science Program to undercut the report's conclusions. The following year, NASA climate scientist James Hansen accused the White House of restricting his communications with the media and censoring his statements about global warming.[10]

The Bush Administration's approach to GHG policy illustrates a significant limitation of the administrative strategy as well as the enhanced role of the federal courts that legislative paralysis invites. The White House suppressed an EPA regulatory finding that CO2 was a pollutant, which was necessary to justify regulation under the Clean Air Act. Moreover, in its 2007 decision in *Massachusetts v. EPA*, the Supreme Court determined that the EPA improperly rejected the state's petition to set CO2 standards under the Clean Air Act. The ruling further required the agency to conduct an endangerment analysis as a first step toward standard-setting, but the administration moved slowly. If, using the best scientific evidence available, the agency determined that CO2 posed a risk to human health it would be obligated to begin rulemaking under section 202 of the Clean Air Act. However, a year passed before the EPA issued an advance notice of proposed rulemaking (ANPR) on greenhouse gasses in July 2008. This stalling effectively punted the issue, leaving the endangerment analysis and the implications of its findings to the next president to address.

The lone regulatory action to address GHG emissions occurred very late during Bush's tenure in office. A 2007 directive to the EPA, Department of Transportation and Department of Energy instructed the three agencies to cut emissions from automobiles and non-road vehicle engines. The directive was followed by an executive order that required the agencies to coordinate with the OMB and specified that the actions would be subject to cost-benefit analysis.[11] Even then, the administration signaled that it would move forward cautiously on GHG regulation. The White House policy establishment remained in conflict not only with the regulators and scientists within its own administration, but also with many state governments, which wanted the right to set their own course.

Regulatory Preemption

As federal regulation of GHG emissions was largely suppressed, there was a burst of innovation at the state level. A decentralized response to climate change was already underway when Bush took office, however. State governments were experimenting with various policies in the 1990s, including greenhouse gas inventories and even CO2 standards for new stationary sources. Barry Rabe attributes this early innovation to a growing consciousness and commitment to address the problem by state governments, as well as to Clinton-era experiments, such as the National Environmental Performance Partnerships, which granted states greater flexibility in exchange for improved environmental performance.[12] Once Bush took office, and it became clear that Kyoto would not be ratified, many states took it upon themselves to reduce GHG emissions. Several states created carbon sequester regulations, and renewable energy mandates became common.[13] By 2012, 37 states and the District of Columbia had enacted their own renewable portfolio requirements.

It was not very long before some states came into direct conflict with the Bush Administration. In 2003, a group of New England states filed suit to force the EPA to regulate CO2 as a pollutant and California began developing CO2 emission standards for mobile sources. As the pressure from the states mounted, the administration pushed back by denying California the waiver it needed to set its own CO_2 standards. It was in this context that the administration began the practice of adding preemptive language to the preambles of regulations, which foreclosed the possibility of states enacting stricter standards than those approved at the federal level.

The conflict over the administration's preemptive federalism boiled over when the Oversight Committee of the U.S. House of Representatives issued a subpoena for communications between the Executive Office of the President and the EPA regarding the waiver denial. The White House continued to tighten its grip on the regulatory state in 2008 when it took the unprecedented step of extending executive privilege to EPA communications.[14] The move was sparked by the EPA's denial of California's request to impose stricter ozone emissions standards and CO_2 standards under the Clean Air Act. After the waiver was rejected, the White House refused a Congressional request for communications between OIRA and the EPA. As House Oversight and Government Reform Committee Chair Henry Waxman protested, "the President has asserted executive privilege to prevent the committee from learning why he and his staff overruled the EPA. There are thousands of internal White House documents that would show whether the president and his staff acted lawfully, but the president has said they must be kept from the public."[15] The waiver was eventually granted, but because the standoff over executive privilege was unresolved, the episode illustrated the power of executive privilege and how easily it could be asserted. This lesson was not lost upon the Obama White House.

The centralized process of regulatory review not only enhanced presidential control over regulation, it contributed to the decline in the independence of regulatory agencies. Another factor was a recently enacted law that aided the Bush administration's business allies. The Data Quality Act of 2000 allowed regulated interests to challenge information used to formulate rules. In effect, this created an additional step in the rulemaking process, allowing parties to question whether rules were indeed formulated on the basis of "sound science."[16] The act accomplished legislatively what prior administrations often did informally: moving the "back door" for business interest appeals out into the open.

Incremental Responses to a Global Threat

Because of the dire state of the economy and the wars Obama inherited, the climate was not a top-tier priority when he became president. Ultimately, he showed a greater commitment to addressing climate change as president than his predecessors, drawing upon all the resources of the office at various times. Applying Soden's typology, we can observe Obama utilizing the roles of party leader, chief diplomat, chief legislator, commander in chief and especially chief executive. As his party's standard-bearer, he set the agenda for his party in Congress and his own administration. Obama took the position that climate change was a serious threat but spoke of it sparingly during the campaign. Initially, the White House strategy was to address climate change indirectly by focusing on energy development and economic recovery, which presented opportunities for green investments.[17]

Obama used his role as legislative leader to press his most ambitious priorities, and he was prolific during his first year in office. One major piece of legislation with significance for the climate was the 2009 Reinvestment and Recovery Act. Promoted as a stimulus for the ailing economy, the act included over $41 billion for energy projects (See Table 6.3). Few of the one-time expenditures in the bill were intended to reduce GHG emissions directly. However, the various conservation, infrastructure, and alternative energy programs were expected to have a significant indirect effect, by cutting down on the amount of fossil fuel consumed. The new administration should have been in a strong position to resume a legislative process after Obama revoked Bush's policy toward federalism with an executive order, permitted the EPA to resume the rulemaking process halted by his predecessor, and even proposed cap-and-trade legislation similar to the bill sponsored by his republican rival for the presidency. Rather, the gridlocked Congress was in no better position to agree upon climate legislation in 2009 than it had been in 1999.

TABLE 6.3 Energy Programs in the American Recovery and Reinvestment Act

	Name of Program
$3,200,000,000	Energy efficiency and conservation block grants
$5,000,000,000	Weatherization Assistance Program (expansion)
$3,100,000,000	State energy program
$2,000,000,000	Advanced batteries manufacturing
$4,400,000,000	Electricity grid modernization
$100,000,000	Electricity grid worker training
$3,400,000,000	Fossil energy research and development
$390,000,000	Uranium Enrichment Decontamination and Decommissioning Fund
$1,600,000,000	Department of Energy science programs
$400,000,000	Advanced Research Projects Agency
$6,000,000,000	Innovative technology loan guarantee program
$10,000,000	Western Area Power Administration construction and maintenance
$3,250,000,000	Bonneville Power Administration borrowing authority
$3,250,000,000	Western Area Power Administration borrowing authority
$500,000,000	Leading edge biofuel projects
$4,500,000,000	Federal building conversion to high-performance green buildings
$300,000,000	Energy efficiency federal vehicle fleet procurement
$41,400,000,000	**Total Energy Programs**

The major piece of legislation intended to address GHG emissions directly was the Waxman-Markey cap-and-trade bill, officially known as the Clean Energy and Security Act of 2009. The bill passed by a narrow margin in the House of Representatives and then gathered dust in the Senate. Supporters hoped that the bill would receive a vote before delegates convened in Copenhagen to ratify a new climate protocol. Earlier in 2009, the EPA announced plans to regulate CO_2 emissions under the authority of the Clean Air Act, supported by the recent Supreme Court decision. The rulemaking effort would be an exercise in futility if Congress created a cap-and-trade market first, but the threat of administrative policymaking can sometimes pressure Congress into acting on a legislative proposal. Illustrating the point, then-Senator John Kerry visited the Copenhagen summit and spoke for the bill's supporters in Congress. Roaming the corridors as if it were the Senate cloak room, he reassured the delegations that there was pressure on Congress to deliver a comprehensive climate bill. "If we don't legislate, they will regulate," echoed Kerry's refrain in speeches before the assembled conferees and smaller audiences.[18]

Once the cap-and-trade bill faltered, climate change was no longer a priority for Congress, an institution populated with members who openly alleged that climate science was a "hoax."[19] By contrast, the executive branch was taking the threat seriously, proceeding with its own mitigation policies and adaptation plans for the changing climate. The mitigation plans take the form of direct regulation, such as emission standards, and indirect regulation in the form of energy efficiency standards. Obama issued Executive Order 13514, which required agencies to consider climate risks and develop plans to help manage the impact of the changing climate upon their operations. It was one of several ways Obama took the initiative to move forward in spite of the paralysis that beset Congress. Every federal agency was required to consider what would happen when the sea levels rise and temperatures are extreme. For example, in 2013, the Internal Revenue Service used data from NASA and a real-world case study of one of its buildings that had suffered flooding in the past in a day-long simulation with other federal agencies. The exercise was designed to address how climate risks might affect the IRS, the first step in preparing a climate preparation and adaptation plan for its own building and surrounding areas.[20] By creating this blueprint, other federal buildings in the Washington, DC area will have a plan to adapt to their own potential vulnerabilities.

Obama took initiative in his capacity as commander-in-chief of the military, issuing environmental directives to the armed forces. For example, he ordered that solar panels be placed on military bases. In 2010, he appeared at Nevada's Nellis Air Force Base, where a 140-acre solar array was under construction.[21] In 2013, the Army announced plans for its largest solar project, a 200-acre array of 94,000 solar panels at Fort Bliss in sunny El Paso, Texas. The base, which already boasted a LEED-certified headquarters building, also had a goal of becoming "Net Zero" by the end of the decade, meaning zero net energy consumption and zero carbon emissions annually.[22] Nancy Sutley, who led the CEQ during Obama's first term, said the Pentagon needed very little persuading because there were obvious benefits of going green:

> It was a whole-of-government approach, where everybody saw themselves as having some environmental responsibility. The biggest change that I saw was in the dept of defense... I think the dept of defense really embraced the environment, climate change, and energy that were not as front and center as fighting two wars, but a much more central issue. They took responsibility for addressing those issues than they might have previously.[23]

As part of the administration's efforts to reduce oil imports, double the production of renewables in three years and develop alternatives such as biofuels, the Navy, DOE and USDA announced a partnership to use and research biofuels. To highlight this partnership, Agriculture Secretary Tom Vilsack stepped aboard the USS *Monterey* to discuss green fleet initiatives and the benefits if buying biofuels in bulk, declaring:

> Developing the next generation of advanced biofuels for our nation's military is both a national security issue and an economic issue.... By utilizing renewable energy produced on American soil, our military forces will become less reliant on fuel that has to be transported long distances and often over supply lines that can be disrupted during times of conflict. Meanwhile, a strong and diverse biofuels industry will support good-paying jobs in rural America that can't be shipped overseas. Through this joint effort, USDA and the U.S. Navy have the opportunity to create a model for American energy security while ensuring the safety of our troops and the long-term viability of our armed forces.[24]

Efforts by the military to rely on less foreign oil had begun under George W. Bush. At the time, the Air Force explored turning coal into liquid jet fuel, but biofuels became the focus under Obama. Currently, F-15 and F-16 fighter as well C-16 transport aircraft have been approved to use up to 50 percent biofuels. Although relying on less foreign petroleum reduces the military's vulnerability to instability in oil producing regions overseas, it also costs more.[25] Skeptics on the House Armed Services Committee, the Senate Subcommittee on Water and Power, and other critics argued that paying nearly four times more for bio fuels is a waste of taxpayer money. Navy Secretary Ray Mabus, a supporter of the Green Fleet initiative, argued that the reliance on foreign fuel sources did not serve the best interests of the Navy. As he argued, a $1 increase in the cost of a barrel of oil adds $31 million to the Navy's tab.[26] Congressional concerns about cost became another obstacle for the administration's agenda.

Executive Leadership and the Climate

Obama's impact was strongest in his role as Chief Executive. Taking the reins of the administrative state, Obama authorized spending and procurement decisions and issued directives to promote energy efficiency and development of renewables. His Executive Order 13514 on Sustainability imposed multiple requirements on the federal government, including planning for a changing climate, such as the IRS exercise described above. In addition to preparedness, E.O. 13514 and Obama's "Greengov" initiative promoted sustainable energy practices amid all federal agencies. Using the federal government as a role model for industry became a primary Obama White House tactic: "As the largest consumer of energy in the U.S. economy, the federal government can and should lead by example when it comes to creating innovative ways to reduce greenhouse gas emissions," Obama announced when the executive order was issued.

EO 13514 directed the newly created Interagency Climate Change Adaptation Task Force to make recommendations, within one year, for how Federal agencies can adapt their policies to take into account a changing climate. A first for the U. S. government, climate change adaptation plans include economic and infrastructure concerns and detail how the government will respond to severe weather, rising sea levels, and other changing climate conditions. The task force also examined ways to protect freshwater resources, fish, wildlife, and plants and consulted with agencies on their plans.

To help the ailing economy and address greenhouse emissions, the White House incorporated a number of green initiatives in its economic stimulus plan that ranged from block grants for energy efficiency projects, to the goal of putting 1 million plug-in hybrid cars on the road by 2015. The plan also called for an $11 billion investment in the so-called "smart-grid." Improving and modernizing the nation's electric grid was a campaign theme for Obama and stimulus spending helped to spur activity. Former EPA Administrator Carol Browner, who served as climate policy advisor during the first two years of the administration, explained the connection between the stimulus and their environmental goals:

> He [Obama] talked about it during the campaign, and I think what we've been able to do in the stimulus is really fill that out. So, as I said, it's bigger, better, smarter. Bigger means we need the new high-voltage lines for renewables. Better means we need to take our existing lines and upgrade them. And then, smarter: We need to make sure that we're really moving electricity in the smartest way and using the most cost-effective electricity at the right time of day. Eventually, we can get to a system where an electric company will be able to hold back some of the power so that maybe your air conditioner won't operate at its peak, you'll still be able to cool your house, but that'll be a savings to the consumer. And so [we will be] giving people and companies a role in the management of how we use electricity.[27]

Just two years after the stimulus money began flowing, Energy Secretary Steven Chu announced that some 5 million smart meters and 140,000 programmable communication thermostats had already been installed nationwide. He underscored the importance of these measures by observing, "To compete in the global economy, we need a modern electricity grid...An upgraded electricity grid will give consumers choices and promote energy savings, increase energy efficiency, and foster the growth of renewable energy resources."[28]

A number of state and local entities benefitted from the grants for smart grid improvements. For example, electric companies in Wyoming received federal grants to help modernize their infrastructure. Overall, the DOE granted some $3.4 billion for power grid modernization projects such as Cheyenne Light, Fuel, and Power's plan to install "smart meters," for which the company received a $5 million grant. The smart meters are a more efficient way to monitor electricity usage.[29] Nearly 90 percent of these funds for smart meters went to blackout-prone communities in Florida, Texas, California, Idaho, Arizona, Oklahoma, Michigan, and Nevada in an effort to prevent blackouts or restore power more quickly. Under Chu, the Department of Energy under Obama expanded the Advanced Research Project Agency-Energy (ARPA-E) to bring together scientists and entrepreneurs to develop new technology for sustainability. ARPA-E was born from George W. Bush's America Competes Act in 2007 and with some $400 million allocated by Congress during the Obama administration in 2009, there have already been nearly 300 projects funded from fuel efficiency to doubling the world-record energy density for a rechargeable lithium-ion battery. For example, some $40 million was set aside for projects that reduce greenhouse emissions from cars and trucks by searching for ways to make vehicles out of lighter materials and fuel them with natural gas that has been converted to liquid fuel.

The largest energy-related item in the Recovery and Reinvestment Act was a $6 billion loan guarantee program for technological innovation. Even though the program was approved by Congress as a one-time expenditure, critics of the stimulus within Congress nonetheless seized upon these grants as evidence of waste. A controversy erupted in 2011, when a solar panel manufacturer called Solyndra defaulted on a $528 million loan. The loan program absorbed the cost of the bankruptcy at no additional cost to taxpayers, but opponents of the program used the controversy to curb future research and development of renewables by the Department of Energy. Harnessing the enormous potential of government procurement is another strategy employed by the Obama White House. The practice is spelled out in the Federal Register, where it explains that agencies should be "pursuing opportunities with vendors and contractors to address and incorporate incentives to reduce greenhouse gas emissions..."[30] Overseeing the implementation of procurement efforts was Jonathan Powers, the Obama Administration's Federal Environmental Executive (FEE) who is tasked with coordinating green government initiatives. The position of FEE was first created by President Clinton, but has been expanded, as Powers recently noted, "...but it's grown. Really, it focused on buying more recycled goods in the beginning [under Clinton]. That was the drive. But as it developed, the executive order has empowered it to do more. A lot of the White House focus was on the broader policy implications."[31]

The Obama White House relied on a number of directives to address global warming. President Obama ordered the Federal Government to reduce its greenhouse gas emissions from direct sources such as building energy use and fuel consumption by 28 percent by 2020. This includes requiring federal agencies to reduce their greenhouse gas emissions from indirect sources, such as those from employee commuting, by 13 percent by 2020. The Obama Administration estimated the federal government would save up to $11 billion dollars in energy costs and the equivalent of 235 million barrels of oil over the next decade.

Additionally, an Obama directive revived plans for an online GHG emissions inventory. The first proposal for an emission registry stalled at OMB in 2008.[32] At the beginning of his first term, Obama restarted work on the database, which tracks greenhouse gas pollution from 29 industrial sources. Similar to the TRI discussed in Chapter 5, this tracking and reporting system promises to become an important tool in the development of a future regulatory regime. Furthermore, supporters of non-regulatory, information disclosure policies suggest that the transparency of a GHG inventory could provide a powerful incentive for polluters to reduce their emissions voluntarily.[33]

An Obama directive created an interagency task force to create government-wide estimates of the benefits of reducing CO2 emissions. The impetus for the directive was a 2007 ruling by the Ninth Circuit Court, which struck auto emission standards as arbitrary and capricious because NHTSA had not justified the benefits. The ruling threatened to unravel rulemaking across the federal government unless NHTSA and other agencies developed a standardized measure that it could defend. After consulting with the EPA and OMB, the task force determined the Social Cost of Carbon to be $21 per metric ton of CO2, an estimate some economists derided as too low.[34]

Modest regulatory changes like the ones described above could have a significant effect on GHG emissions even in the absence of climate legislation. Ironically, by asserting its authority to act on GHG emissions in the face of Congressional paralysis, the EPA may have made legislation even less likely. This willing regulatory activism takes the pressure off Congress, and this puts the heat on the bureaucracy for making the decisions that members of Congress have steadfastly avoided. On the eve of the EPA's 2009 announcement of an endangerment finding on CO2, an industry lobbyist said in an interview, "The Obama Administration now owns this political hot potato," and asking, rhetorically, why Congress would ever "want to take it back?"[35]

Environmental Diplomacy

Every recent president has had to engage the issue of the climate from the role of chief diplomat. Negotiating any international agreements, let alone climate treaties, on behalf of the U.S. is an unenviable chore, considering the arduous ratification process. Nevertheless, parties at the 2009 U.N. Climate Change Summit in Copenhagen believed President Obama's presence was requisite to reaching any agreement, even though representatives of the State Department, including Secretary of State Hillary Clinton, had been present carrying out negotiations.

By Obama's first year in office, the conditions were ripe for a breakthrough on an enforceable worldwide climate agreement. Representatives from 190 nations planned to meet in Copenhagen to negotiate a treaty to make the targets from Kyoto binding. In a sign of the urgency the surrounded the talks, 115 heads of state were among the delegations.[36] The major domestic legislation that Obama needed to demonstrate U.S. commitment seemed within reach a few weeks earlier, but it was already slipping away by the start of the summit in December.

The Senate had still not scheduled a vote on the climate bill by the time the meetings convened. Administration officials began to downplay the significance of the delay, and suggested that Obama could skip the meeting altogether. Instead, they put a positive face on the development and made a display of the administration's resolve to pursue climate policies with or without the stalled cap-and-trade legislation. The U.S. delegation included no fewer than nine cabinet and cabinet-level officials. Among them were the EPA Administrator Lisa Jackson, CEQ Chair Nancy Sutley, Energy and Climate Advisor Carol Browner, Science Advisor John Holdren, and five cabinet secretaries. They presented detailed plans to use the authority of their own agencies to address climate change in specific ways, while reinforcing a shared theme that the administration does not need Congress to act.

Jackson arrived at the opening of the summit just one day after announcing the EPA's endangerment finding that declared CO_2 a pollutant. She delivered a message to reassure the parties, explaining that the finding removed any doubt that the agency had the authority to regulate CO_2:

> This administration will not ignore science or the law any longer, nor will we avoid the responsibility we owe to our children and grandchildren. By taking action and finalizing the endangerment finding on greenhouse gas pollution, we have been authorized to and obligated to take reasonable efforts to reduce greenhouse pollutants under the Clean Air Act.[37]

The implied message was that a reduction in emissions could be mandated with or without new legislation. Interior Secretary Ken Salazar addressed the delegates on the second day. He took a shot at former President George W. Bush, asserting, "As much of the world awoke to the dangers of climate change, the political leadership of the United States simply slept." Salazar's goal was to reassure delegates that the new administration was rolling back Bush-era policies, as well as promote the Interior Department's own initiatives, notably the leasing of public lands for wind and solar projects. He was followed on the third day by Commerce Secretary Gary Locke, who touted green jobs created as a result of the stimulus bill.

Other cabinet secretaries boasted about the projected impacts of the initiatives already underway. Agriculture Secretary Tom Vilsack used the conference as an occasion to announce a public-private partnership with dairy farmers to reduce the industry's methane emissions by 25 percent by 2020 by making anaerobic digesters available. Energy Secretary Steven Chu spoke about the benefits of new energy efficiency standards the Energy Department was promulgating for more consumer appliances. Combined, the standards were predicted to save 18.4 billion kilowatt hours annually by 2020, or the equivalent of five coal-fired power plants running at full power for an entire year. Chu was also the first to announce the U.S. pledge of $350 million for a clean technology fund to aid developing countries. This became the foundation for the UN Green Climate Fund, which grew to more than $10 billion in commitments from industrialized countries by 2017.

U.S. cap-and-trade legislation notwithstanding, the Copenhagen talks were ultimately doomed because of the resistance of China and India to the proposed targets and monitoring regimes. Near the end of the conference, another representative of the administration worked to lower expectations. The most important thing, according to Obama's science advisor John Holdren, was for U.S. emissions to level off and begin to decline before 2020. Since the U.S. was on already track to do this under current initiatives, he suggested it would become much easier to commit to targets and adjust the rate of change in the future.[38]

Obama's arrival on the final day of the summit was hotly anticipated. The president brought nothing new to the negotiations, but his remarks summarized the commitments already announced by his administration and his presence underlined the administration's commitment. Individually, each spoke of administrative actions that could have been construed as symbolic gestures. However, the combined initiatives offered the potential for significant unilateral reductions in CO_2 and other GHG emissions. Obama argued that his administration was making significant progress unilaterally, by making bilateral agreements and international partnerships, and progress on GHG reductions as a benefit of the Montreal Accord. The U.S.-backed climate mitigation fund was a significant commitment that emerged from the conference, as well as an agreement to protect the world's forests.

Negotiations ultimately broke down, as parties rejected enforceable targets in favor of a nonbinding resolution to limit global warming by two degrees. Blame for the collapse was spread around several of the parties, but some have suggested the process was fatally flawed. In a critique of the U.N. approach, David Victor argues that the process and set up unreasonable expectations for quick achievement of largely symbolic goals (such as the two-degree target) with too little attention to helping various countries find practical approaches to reducing emissions that suited their own circumstances. Ultimately the Copenhagen talks were destined to be gridlocked because too many parties were involved in the process too early.[39]

Paris and Beyond

Even before the Waxman-Markey stalled in the senate, advisors had suggested that the EPA could regulate carbon emissions or create a market under its own authority. The Supreme Court noted this solution was available in its recent decision in *Massachusetts v. EPA*. President Obama, who had spoken about using executive authority in the face of Congressional inaction, faced increasing pressure to do just that.

In the State of the Union address that kicked off his second term, President Obama encouraged Congress, "We are not rivals for power, but partners in progress." That quotation from John F. Kennedy was originally a powerful appeal not for bipartisanship, but rather for cooperation between the branches of government in pursuit of common goals. Obama's conciliatory words notwithstanding, few of the objectives he proposed at the beginning of his second term were expected to advance smoothly through the legislative process. The proposals put forth in this opening address seemed less of an agenda for Congress than an effort to stake out positions and rally the public. He touted successful public investments in research and development for renewable energy during his first term, spending that was now on the chopping block. Then the speech turned to the environment, demanding that the federal government do more to combat climate change. This time, the message urged cooperation across both party and branch lines:

> I urge this Congress to pursue a bipartisan, market-based solution to climate change, like the one John McCain and Joe Lieberman worked on together a few years ago. But if Congress won't act soon to protect future generations, I will. I will direct my Cabinet to come up with executive actions we can take, now and in the future, to reduce pollution, prepare our communities for the consequences of climate change, and speed the transition to more sustainable sources of energy.

The tone was as defiant as it was urgent. Asserting that he had sufficient resources within the executive branch as well as a moral imperative to use them, Obama seemed to dare his opponents in Congress to try to stop him.

For environmentalists, it was a welcome display of resolve. With Obama's first election four years earlier, they were hopeful that their priorities would again enjoy Washington's favor and support. For eight years, the Bush Administration had prioritized fossil fuel production over conservation and had scaled back pollution policies in deference to industry. In sharp contrast, Obama's ambitious agenda included several environmental policy goals. However, through the first term, Obama's efforts to achieve them would be modest, and he cautioned that the environmental community should be modest with their expectations. First, his early statements about the environment were usually coupled with expressions of support for environmental growth. Second, he indicated that he intended to bypass Congress and make use of administrative policy instruments, relying on executive rulemaking to advance certain environmental goals. Despite this willingness to act unilaterally, this also implied that he intended to conserve his political capital with Congress for other battles over health care and financial reform.

There was little hope for cooperation with Obama's agenda in the 112th Congress. Most of the Democrats defeated in the 2010 midterm elections were moderates, unlike the freshman Republican class. Therefore, the House of Representatives became the most conservative and ideologically polarized House in the modern era.[40] Obama managed to pass health care reform legislation (The Affordable Care Act of 2010) when Congress still had large Democratic majorities. The midterm election put the brakes on other initiatives, such as comprehensive immigration reform and climate change legislation, which were deferred when he essentially "bet his domestic presidency on health care."[41] Just as his immediate predecessors had done, Obama would be content to govern largely without Congress. Contrary to the conciliatory State of the Union, the president and Congress would remain rivals, rather than build an effective partnership.

Obama's second-term regulatory offensive on GHG emissions would be spearheaded by EPA Administrator Gina McCarthy, who previously served as the agency's chief enforcement officer under Jackson. Under her watch, the EPA finalized standards to double the fuel economy of new automobile fleets by 2025 reducing their GHG emissions by 3.7 percent annually. The most consequential policy promulgated during Obama's second term was the Clean Power Plan. The regulations were designed to cut GHG emissions from existing power plants 30 percent below 2005 levels by 2030. Under the policy, the states would submit their own plans to meet their GHG budgets, which were based on their previous emission levels, and determine the best way to meet their reduction targets. Rather than set technology-forcing standards for power plants, the plan gave each state the discretion to meet the targets in a way that suited their own needs.

The plan became a target for Congressional Republicans. Multiple House and Senate committees held hearings on the plan, and Senate Majority Leader Mitch McConnell actively discouraged states from complying with the policy, calling it legally questionable.[42] Nonetheless, the Clean Power Plan set a timetable for the US power sector on a course to reduce carbon emissions for the first time. This allowed Obama to secure a bilateral agreement with China, the largest source of GHG pollution, to commit to its own emissions targets. In November 2014, Obama and Chinese President Xi Jinping announced the historic agreement in a joint news conference. Both leaders agreed climate change was one of the most serious threats facing humanity, promised to pursue domestic climate policies, and proclaimed their shared commitment to a successful global agreement in the next round of UN negotiations the following year.[43] Because China and the US had been the two most defiant parties at Copenhagen, this agreement represented the most significant diplomatic breakthrough in climate negotiations outside of the UN Framework.

Having secured China's cooperation, Obama attended the 2015 UN Climate Summit in Paris in a stronger position compared to Copenhagen six years earlier. It was understood that the US would not agree to a binding, top-down emissions reductions plan, so the parties took a different approach. The Paris Agreement centered on five-year cycles, in which each country would determine its own preferred method of meeting its nationally determined contribution (NDC) of carbon emissions. This was designed to meet the agreed-upon limit of two degrees of warming, and further emissions reduction in subsequent cycles was expected hold warming to 1.5 degrees above preindustrial levels. In the US, such progress was now possible because of the EPA's mandate to regulate CO_2 emissions following *Mass v. EPA*, and if necessary, this could be accomplished without legislation.

The Paris Agreement was generally well-received because it relied upon individual countries to make their own implementation decisions. This flexibility was one reason for its success. Another factor was Obama's strategy of pursuing a policy that did not require ratification by the senate. Unlike the Kyoto Protocol and the Waxman-Markey Bill, which was intended to implement the plan proposed at Copenhagen, the Paris Agreement was not contingent on Congressional approval. It was an executive agreement built upon on the success of his prior executive agreement with China.

Toward the end of his second term, Obama was still under pressure from the environmental community to show results. The failure of the Waxman-Markey bill was a bitter defeat. Subsequently, he scored significant policy accomplishments with his use of presidential directives, executive agreements, spending and procurement, but these accomplishments were the products of behind-the-scenes executive actions and negotiations. One issue which took on outsized importance was approval of the Keystone XL pipeline, which was proposed to transport tar sand bitumen from Alberta to the refineries and ports on the Gulf of Mexico. The pipeline quickly became a symbolic issue. Few people outside of government and the petroleum industry had even heard of the proposed pipeline before the environmental community took it up as a major cause. In an op-ed, James Hansen wrote that if the pipeline were completed, it would be "game over" for the climate. The international pipeline required State Department approval, and Secretary of State John Kerry ultimately decided not to let the project proceed. As with Obama's other executive actions on the climate, the decision to deny the Keystone XL Pipeline was vulnerable to reversal by the following administration. This is precisely what happened when Donald Trump unexpectedly won the 2016 election and named former Exxon-Mobil Chairman Rex Tillerson his first Secretary of State.

Climate Policy Retrenchment

The rollback began immediately after Trump's inauguration. In no area was this campaign of retrenchment more aggressive than in climate policy. These were done on behalf of a handful of privately-owned coal and petroleum companies, which were among the few businesses to support his campaign at the beginning. It is not uncommon for the interests to seek favors from the politicians they help elect, but they tend not to be transparent about this to avoid the appearance of a quid pro quo. One of Trump's early and influential supporters was Robert Murray, who donated $300,000 to his inauguration committee and whose associates landed key roles in Trump's transition team and EPA (see Chapter 2). Murray had access and influence, and he did not hesitate to provide the newly elected president with a "wish list." In a March 1, 2017 letter, Murray listed 16 "actions in order of priority."[44] The policy changes Murray outlined are listed in Table 6.4. The list was comprised mostly of actions that could be taken with the president's authority as chief executive, including regulatory rollbacks, cuts to the EPA and Mine Health and Safety Administration, and appointments to the Supreme Court, of Federal Energy Regulatory Commission, and National Labor Relations Board. Only one of the 16 action items (public funding for mine worker pensions) would have required legislation, suggesting this coal baron looked to the administration as the institution that would deliver results.

TABLE 6.4 Action Plan for the Administration of President Donald Trump

Rescind the Clean Power Plan

Withdraw the EPA's Endangerment Finding on Greenhouse Gasses

Eliminate the 30 percent tax credit for wind turbines and solar panels

Withdraw from the Paris Climate Accord

Repeal Ozone Maximum Achievable Technology Standards for electric utilities

Increase funding for clean coal technologies

Reduce the number of coal mine inspectors at MSHA

Cut the staff of the EPA by 50 percent

Overturn the Cross-state Air Pollution Rule

Revise the MSHA's Coal Mine Dust Rule

Fund pensions for retired coal miners

Rescind the MSHA's Pattern of Violations Rule

Appoint conservative justices to the Supreme Court

Replace the members of Federal Energy Regulatory Commission

Replace the members of the Tennessee Valley Authority Board of Directors

Replace the members of National Labor Relations Board

Within two weeks of Murray's letter, Trump signed an executive order representing a multi-pronged attack on Obama's climate legacy. Executive Order 13783 dissolved the inter-agency working group on the social cost of carbon (SCC).[45] Federal agencies were instructed to abandon the administration-wide SCC, and begin conducting their own cost-benefit analyses for regulations using new calculations which considered only the social cost for the US economy, rather than the global economy. As the following chapter will show, this has profound implications for policies designed to limit climate change. One section of EO 13783, which received far more media attention at the time, announced Trump's intention to repeal the Clean Power Plan. This would require the EPA to undertake a two-year rulemaking process. Its replacement, known as the Affordable Clean Energy (ACE) rule, directs the states to reduce GHG emissions but sets no targets. This scaled back the EPA's role in setting CO2 standards, largely giving the states the authority to decide how much to reduce emissions, if at all. The rule was subject to multiple legal challenges, with six states announcing a lawsuit to block the measure.[46]

Trump also suspended the revised tailpipe emission standards for new automobile fleets, even though manufacturers already planned to comply with the new regulations. Subsequently, the state of California reached a deal with four major automakers to boost the average fuel economy of their fleets to 50 miles per gallon—the same goal as the suspended Obama-era regulation—by 2026. Because of the size of the state's market, and the demands from European and other foreign markets for cleaner vehicles, California's action makes it more likely that its standard will become the de facto national policy.[47]

In May 2017, Trump announced the US would begin the process of withdrawing from the Paris Agreement. In a Rose Garden appearance, Trump said, "We are getting out... but we will start to negotiate, and we will see if we can make a deal that's fair."[48] He was unable to leave unilaterally because as the agreement was worded, no party could withdraw for at least five years and without giving one year's notice. According to former State Department Counsel Harold Hongju Koh, this provision was added at the last minute just in case the winner of the 2016 presidential election attempted to withdraw.[49] In November 2019, the White House announced the US would exit the agreement after the November 2020 election, which was the earliest possible exit date. As a multi-lateral agreement, the agreement cannot be renegotiated by one country, so Trump's unilateral move perplexed the other parties as much as it angered them. Other world leaders, as well as US governors and mayors responded by declaring they would continue to implement the agreement. At the same White House ceremony, Trump also had harsh words for the Green Climate Fund and criticized Obama's pledge to contribute $3 billion the previous year.

In another gesture for the industry, Energy Secretary Rick Perry invoked national security in a proposal to subsidize coal (and nuclear) power plants. The rationale for his plan was that subsidies should be given to power stations that maintained a 60-day supply of fuel on site, as a precaution against an emergency that might cause a disruption in the supply chain. Renewable energy sources would be at a disadvantage under such a policy, because of the storage issues associated with wind and solar energy. However, the proposal divided the fossil fuel industry, because it made natural gas less competitive as well. Perry ultimately withdrew the proposal in the face of opposition within the administration, as the justification for the plan was rejected by both the National Security Council and the National Economic Council.

Politicizing Climate Science

Trump's strategy of deconstructing the administrative state went beyond suspending new regulations. In order to pursue this retrenchment agenda, the White House would have to disregard or discredit the research and analysis behind the climate policies enacted by his predecessor. In an early sign of the administration's hostility toward science, the EPA transition team included Steven Milloy, a former lobbyist for the tobacco and oil industries, and Myron Ebell, who previously provided research for the tobacco industry and was now focused on discrediting climate science. When a member of the Trump transition staff first visited the Department of Agriculture a month after the election, it was reported that he was preoccupied with finding out how many people in the department did research on climate change and what their names were.[50]

Within days of Trump's inauguration, observers noticed that mentions of climate change had been removed from federal agency websites. Even the EPA's "Climate and Energy Resources" web page was scrubbed of mentions of climate change.[51] From other agencies, there were similar stories. Employees of the Center for Disease Control complained Trump appointees had banned them from using certain phrases, including "evidence-based" and "science-based" in their analyses.[52] Later, in a trial balloon, the acting director of NOAA proposed the agency remove the word "climate" from the agency's mission statement.[53]

Scientists in the administration perceived harassment by the political appointees and the White House. At the EPA, a political appointee was placed in charge of vetting research grants. John Konkus, who officially worked in the agency's public affairs office, reportedly told staff he was on the lookout for "the double C-word," and in 2017 cancelled more than $2 million in grants that had already been awarded to universities and nonprofits.[54] Moreover, several academic scientists were dismissed from the agency's advisory boards and replaced with industry representatives. The Government Accountability Office later determined these firings and replacements, initiated by Administrator Scott Pruitt, violated federal ethics rules.[55] Scientists working elsewhere within government also found their actions restricted and their work censored. For example, the US Geological Survey was ordered to stop its climate projections in 2040 instead of 2100, as its practice had been. A State Department analyst resigned after accusing the White House of blocking testimony he had planned to present to Congress. The analyst, Rod

Schoonover, planned to tell the House Intelligence Committee, "Climate change will have wide-ranging implications for US national security over the next 20 years through global perturbations, increased risk of political instability, heightened tensions between countries for resources, a growing number of climate-linked humanitarian crises....". The White House Office of Legislative Affairs ultimately decided to allow Schoonover to testify before the committee, but he was not allowed to submit his written statement.[56] Maria Caffrey, a climate scientist employed by the National Park Service, filed a whistleblower complaint against the Trump Administration after she lost her job. Caffrey claimed she was fired after she submitted a report on the effects of sea level rise on national parks, but its publication was delayed by the NPS and she was coerced into removing references to human-caused climate change.[57]

The assault on climate science served a larger purpose. It cast doubt on the assumptions behind GHG emissions policies and undermined the regulatory state's ability to pursue stricter ones. This included changing the cost-benefit calculations which provided justification for those policies. Based on an exhaustive review of scientific literature, Obama's inter-agency task force determined the SCC was $42 per ton. It arrived at this figure by considering the global effects of climate change. In contrast, the Trump Administration took an "America First" approach, prohibiting agencies from considering the impact of climate change outside the US.

Federal agencies were to consider only the costs of climate change to the US economy. Using these assumptions, the Trump Administration estimated the SCC was closer to $1 per ton. Consequently, the Clean Power Plan could not be justified in terms of cost-benefit analysis.

The Trump EPA proposed a replacement rule, the American Clean Energy (ACE) regulation, which was based on voluntary GHG reductions. Replacing the Clean Power Plan with the ACE regulation was expected to have significant health and mortality effects in the U.S. for reasons other than climate change. An EPA assessment of the original Clean Power Plan projected it would also decrease PM-2.5 pollution enough to prevent 1,400 premature deaths annually by 2030, as well as 15,000 cases of upper respiratory disorders. In its analysis of the ACE regulation, the EPA changed the way it calculated the health risks of pollution. It was assumed that the existing safety threshold for PM-2.5 of 12 micrograms per cubic meter was safe enough, and no additional health benefits could be achieved from further reductions.[58] In effect, the health benefit of further pollution reduction was assumed to be zero, removing the remaining justification for the Clean Power Plan.

Conclusion

Climate change represents an entire area of public policy derailed by a deadlocked legislative process. After Bill Clinton's legislative proposals were blocked and the Kyoto process rejected, he was only successful with voluntary partnerships for energy efficiency and GHG reductions, while initiating regulatory actions that were blocked by his successor. Responding to the threat of climate change was not a priority for George W. Bush, who returned to Reagan's model of an administrative presidency to restrain the regulatory state and promote fossil fuel production. Using administrative tools developed by Reagan, George H.W. Bush and Clinton, Bush prevailed—in the short term—in removing regulatory obstacles for energy production, while containing the environmental bureaucracy and preempting state-level efforts to reduce emissions. The Obama Administration has returned to the model of presidential administration established by Bill Clinton, initiating a large number of modest initiatives under his executive authority, and aligning himself with agency managers and missions.

Because of the effective way Congress represents geographic interests, states and communities that have a stake in fossil fuel production (especially coal) can counter proponents of alternative energy and conservation, perpetuating a policy stalemate. The lack of legislative leadership has opened the way for an active bureaucracy and presidential dominance. The executive branch, with its capacity to move decisively, has most often taken the lead. Bolstered by the Supreme Court's decision in *Massachusetts v. EPA* and other favorable federal court decisions, the Obama Administration was in a stronger position to move forward with the Clean Power Plan to limit GHG emissions.

One consequence of increased regulatory activism by the executive branch is that Congress may be less willing to take responsibility for solving long-term problems of energy independence and global cooperation on the climate, for which Congressional authorization is critical. These cannot be addressed through minor regulatory changes or decisions in spending and procurement. Obama was able to expand the toolkit of the chief executive and address climate policy through directives, spending and procurement. These were modest initiatives compared to the bold action that environmentalists sought. However, because of these regulations, he was able to negotiate a bilateral agreement with China, which set the stage for a global agreement to limit GHG emissions in 2015. Thus, Obama succeeded where his predecessors failed. Because Congress was on the sidelines, these agreements lacked buy-in as well as the legitimacy of a formal treaty.

As with his record on natural resources and pollution policy, Obama's climate policies were limited and provisional. Relying on administrative tools to advance his agenda instead of legislation, most of his accomplishments were vulnerable to reversal by his successor. The Trump Administration took advantage of rule-making, executive orders, and appointments to undo much of this legacy. As the following chapter argues, Trump's impact could be long-lasting, as he did more than his predecessors to politicize science and policy analysis.

The first wave of Trump's deregulation initiatives largely failed. Nearly every effort to block, delay or rescind the rules promulgated through the regulatory process was subject to litigation, and was struck down by the courts. They were mainly successful as symbolic gestures for the benefit of the coal and petroleum interests that supported the campaign. Even though some coal companies and utilities benefited from Trump's deregulation in the short term, the downward slide of the coal industry continued. In 2010, there were 580 coal-fired power plants providing 45 percent of US electric power. By 2018, there were fewer than 350, generating 28 percent. The entire coal industry employed 53,000, in 2018, only slightly more than 20 percent employed by the solar power industry.

After Trump's first two years in office, there was a second wave of policy retrenchment, the impact of which may outlast the administration itself. When he disbanded the interagency working group which set the social cost of carbon, he set the stage for a more sophisticated approach to deregulation. A traditional instrument of presidential administration—cost-benefit analysis—could now be employed selectively and oriented toward justifying certain results. As the following chapter argues, the administration was able to undermine a number of environmental regulations by changing the assumptions used in the implementation process. Thus, the White House determined it could accomplish much of its deregulation agenda not by changing the law or rescinding regulations, but by working within the rules.

Notes

1 U.S. Global Change Research Program, "Fourth National Climate Assessment," https://nca2018.globalchange.gov.
2 The Climate Assessment would have been released in December, to coincide with a scientific conference in Washington, DC.
3 Coral Davenport, 2018, "Trump Administration's Strategy on Climate: Try to Bury its Own Scientific Report," *NYT* (November 25).
4 Lyndon B. Johnson, 1965, "Special Message to the Congress on Conservation and Restoration of Natural Beauty," February 8, www.lbjlib.utexas.edu/johnson/archives.hom/speeches.hom/650208.asp.
5 Harriet Bulkeley and Peter Newell, 2010, *Governing Climate Change*, New York: Routledge.
6 Bill McKibben, 1989, *The End of Nature*, New York: Random House.
7 William J. Clinton, Remarks for Climate Change Event at the White House (July 23, 1997), Clinton Presidential Library: Files-Weiss, Box 7 (Climate).
8 This metric is misleading because carbon intensity was already declining due to conservation and improved efficiencies in technology. See Marc A. Eisner, 2007, *Governing the Environment*, Boulder, CO: Lynne Rienner.
9 Christine Todd Whitman, 2004, *It's My Party Too*, New York: Penguin.
10 Hansen was already well-known for being outspoken. During the Presidency of Bush's father, he accused the White House of altering his testimony to a Congressional panel in which he stated there was "99 percent certainty" that global warming was man-made. See Christopher Mooney, 2005, 2006, *The Republican War on Science*, New York: Basic Books.
11 George W. Bush, 2007, Executive Order 13432—Cooperation Among Agencies in Protecting the Environment with Respect to Greenhouse Gas Emissions from Motor Vehicles, Nonroad Vehicles, and Nonroad Engines Online by Gerhard Peters and John T. Woolley, The American Presidency Project, www.presidency.ucsb.edu/node/275130.
12 Not all of this the policy innovation was aimed at reducing emissions. After the senate passed the Byrd-Hagel Resolution, coal and oil producing states passed their own resolutions critical of Kyoto. Barry Rabe, 2004, *Statehouse and Greenhouse*, Washington, DC: Brookings.

13 Ibid.
14 Mark J. Rozell and Mitchel A. Sollenberger, 2009, "Executive Privilege and the Unitary Executive Theory in the George W. Bush Administration" in *Rivals for Power: Presidential-Congressional Relations*, edited by James A. Thurber, Lanham, MD: Rowman and Littlefield.
15 Henry Waxman, 2008, Opening Statement of House Committee on Oversight and Government Reform, Business Meeting Regarding Contempt Resolution (June 20) One Hundred Tenth Congress.
16 Richard N. L. Andrews, 2006, "Risk-Based Decision-Making: Policy, Science and Politics" in *Environmental Policy: New Directions for the Twenty-First Century*, 6th edition, edited by Norman J. Vig and Michael E. Kraft, Washington, DC: Congressional Quarterly Press, 225.
17 Jonathan Alter, 2010, *The Promise: President Obama's First Year*, New York: Simon and Schuster.
18 David M. Shafie, 2011, "The Presidency and Domestic Policy" in *New Directions in the American Presidency*, edited by Lori Cox Han, New York: Routledge, 166–79.
19 James Inhofe, 2003, Floor speech on "The Science of Climate Change," July 28.
20 U.S. Environmental Protection Agency, n.d., "Federal Green Challenge," www.epa.gov/Region09/federalgreenchallenge/spotlight/capital.html.
21 "Air Force to Quadruple Solar Energy Production" (October 15, 2010). Inside AF: www.af.mil/News/Article-Display/Article/115298/air-force-to-quadruple-solar-energy-production/.
22 Cindy Ramirez, 2013, "94,000 Solar Panels to Help Power Part of Fort Bliss," *El Paso Times*, April 6.
23 Personal communication with author, June 18, 2018, Los Angeles, CA.
24 U.S. Department of Agriculture (USDA), 2012, "Vilsack, Mabus Highlight Navy Energy Independence in Norfolk," News Release, September 10.
25 Elizabeth Shogren, 2011, "Air Force and Navy Turn to Biofuels," NPR.org, September 22.
26 Noah Shachtman, 2012, "Republicans Order Navy to Quit Buying Biofuel," *Wired*, May 14.
27 Kenneth Walsh, 2009, "Carol Browner on Climate Change," *US News* (March 9), www.usnews.com/news/energy/articles/2009/03/09/on-climate-change-the-science-has-just-become-incredibly-clear.
28 U.S. Department of Energy (DOE), 2011, "Energy Secretary Chu Announces Five Million Smart Meters Installed Nationwide as Part of Grid Modernization Effort," News Release, June 13, http://energy.gov/articles/energy-secretary-chu -announces-five-million-smart-meters-installed-nationwide-part-grid.
29 Jeremy Pelzer, 2009, "Smart-Grid Stimulus Funds Come to Wyo," *Casper Star-Tribune* (October 27).
30 Federal Register 74(194):52118 (Thursday, October 8, 2009).
31 Greg Korte, 2013, "Obama's Point Man on the Environment," *USA Today*, April 2.
32 Juliet Eilperin, 2009, "Winds of Change Evident in U.S. Environmental Policy," *Washington Post* (March 30): A3.
33 Richard H. Thaler and Cass R. Sunstein, 2008, 2009, *Nudge: Improving Decisions about Health, Wealth and Happiness*, New York: Penguin.
34 In their critique of the SCC, Johnson and Hope argue that the government used an inappropriate discount rate, based on assumption that poorer regions would have the same ability to cope with the effects of climate change as people in wealthier regions. See Laurie T. Johnson and Chris Hope, 2012, "The Social Cost of Carbon in U.S. Regulatory Impact Analyses: An Introduction and Critique," *Journal of Environmental Studies and Science* 2: 205–221.
35 Kim Strassel, 2009, "The EPA's Carbon Bomb Fizzles," *Wall Street Journal* (December 11).

36 Allegra Stratton and John Vidal, 2009, "Hedegaard Resigns as President of Copenhagen Climate Summit," *Guardian*, December 16.

37 USEPA, Administrator Lisa P. Jackson, Remarks to the United Nations Climate Change Conference in Copenhagen (December 9, 2009).

38 John Holdren, Personal communication with author, December 17, 2009, UN Climate Change Conference, Copenhagen, Denmark.

39 A possibly more fruitful approach has been proposed by David Victor. The plan would involve agreements among a more exclusive "club" of wealthier industrialized nations that could gradually be enlarged. See David Victor, 2011, *Global Warming Gridlock*, Cambridge: Cambridge University Press.

40 Alan I. Abramowitz, 2011, "Expect Confrontation, Not Compromise: The 112th House of Representatives Is Likely to Be the Most Conservative and Polarized House in the Modern Era," *PS: Political Science and Politics* 44(2): 293–296.

41 Jonathan Alter, 2010, *The Promise: President Obama's First Year*, New York: Simon and Schuster.

42 Mitch McConnell, 2015, "States Should Reject Obama Mandate," *Lexington Herald-Leader* (March 3).

43 The White House, 2015, "US-China Joint Presidential Statement on Climate Change," News Release (September 25).

44 Robert Murray, 2017, Letter to Vice President Michael Pence (March 1), www.nytimes.com/interactive/2018/01/09/climate/document-Murray-Energy-Action-Plan.html.

45 Donald J. Trump, 2017, Executive Order 13783, "Promoting Energy Independence and Economic Growth" (March 28), www.whitehouse.gov/presidential-actions/presidential-executive-order-promoting-energy-independence-economic-growth/.

46 The initial suits were filed by the attorneys general of California, Colorado, Iowa, Oregon, New York and Washington. See Lisa Friedman, 2019b, "EPA Plans to Get Thousands of Pollution Deaths Off the Books by Changing its Math," *New York Times* (May 20).

47 Paul Rogers, 2019, "How California Has Blocked Trump's Environmental Rollbacks," *San Jose Mercury News* (July 27).

48 Quoted in Somini Sengupta, 2017, "As Trump Exits Climate Agreement, Other Nations are Defiant," *New York Times* (June 1).

49 Harold Hongju Koh, Personal communication with author, Orange, CA, October 24, 2019.

50 Michael Lewis, 2018, *The Fifth Risk*, New York: Norton.

51 Lisa Friedman, 2017, "EPA Scrubs Climate Website of 'Climate Change,'" *New York Times* (October 20).

52 Christopher Mooney, 2017, "CDC Gets List of 'Forbidden Words,'" *Washington Post* (June 17).

53 Christopher Mooney and Jason Samenow, 2017, "Ocean Science Agency Chief Floats Removing 'Climate' from Mission Statement," *Washington Post* (June 25).

54 Juliet Eilperin, 2017, "EPA Now Requires Political Aide's Sign-off for Agency Awards, Grant Applications," Washington Post (September 4).

55 Lisa Friedman, 2019(a), "EPA Broke Rules in Shake-Up of Science Panels, Federal Watchdog Says," *New York Times* (July 16). Lisa Friedman, 2019(b), "EPA Plans to Get Thousands of Pollution Deaths Off the Books by Changing its Math," *New York Times* (May 20).

56 Rod Schoonover, 2019, Op-ed: "The White House Blocked My Report on Climate Change," *New York Times* (July 30).

57 Maria Caffrey, 2019, "I was a Climate Scientist in a Climate-Denying Administration – and it Cost Me My Job," *The Guardian* (July 25).

58 Lisa Friedman, 2019(b), "EPA Plans to Get Thousands of Pollution Deaths Off the Books by Changing its Math," *New York Times* (May 20).

Bibliography

"Air Force to Quadruple Solar Energy Production" (October 15, 2010). Inside AF: www.af.mil/News/Article-Display/Article/115298/air-force-to-quadruple-solar-energy-production/.

Abramowitz, Alan I. 2011. "Expect Confrontation, Not Compromise: The 112th House of Representatives Is Likely to Be the Most Conservative and Polarized House in the Modern Era." *PS: Political Science and Politics* 44(2): 293–296.

Alter, Jonathan. 2010. *The Promise: President Obama's First Year.* New York: Simon and Schuster.

Andrews, Richard N. L. 2006. "Risk-Based Decision-Making: Policy, Science and Politics" in *Environmental Policy: New Directions for the Twenty-First Century*, 6th edition, edited by Norman J. Vig and Michael E. Kraft. Washington, DC: Congressional Quarterly Press. 225.

Bulkeley, Harriet and Peter Newell. 2010. *Governing Climate Change.* New York: Routledge.

Bush, George W. 2007. "Executive Order 13432—Cooperation Among Agencies in Protecting the Environment with Respect to Greenhouse Gas Emissions from Motor Vehicles, Nonroad Vehicles, and Nonroad Engines." Online by Gerhard Peters and John T. Woolley, *The American Presidency Project.* www.presidency.ucsb.edu/node/275130.

Caffrey, Maria 2019. "I was a Climate Scientist in a Climate-Denying Administration – and it Cost Me My Job." *The Guardian* (July 25).

Clinton, William J. 1997. Remarks for Climate Change Event at the White House (July 23). Clinton Presidential Library: Files-Weiss, Box 7 (Climate).

Davenport, Coral. 2018. "Trump Administration's Strategy on Climate: Try to Bury its Own Scientific Report." *New York Times* (November 25).

Eilperin, Juliet. 2009. "Winds of Change Evident in U.S. Environmental Policy." *Washington Post* (March 30): A3.

Eisner, Marc A. 2007. *Governing the Environment.* Boulder, CO: Lynne Rienner.

Federal Register. 74(194): 52118 (Thursday, October 8, 2009).

Friedman, Lisa. 2017. "EPA Scrubs Climate Website of 'Climate Change.'" *New York Times* (October 20).

Friedman, Lisa. 2019a. "EPA Broke Rules in Shake-Up of Science Panels, Federal Watchdog Says." *New York Times* (July 16).

Friedman, Lisa. 2019b. "EPA Plans to Get Thousands of Pollution Deaths Off the Books by Changing its Math." *New York Times* (May 20).

Inhofe, James. 2003. Floor speech on "The Science of Climate Change." July 28.

Johnson, Laurie T. and Chris Hope. 2012. "The Social Cost of Carbon in U.S. Regulatory Impact Analyses: An Introduction and Critique." *Journal of Environmental Studies and Science* 2: 205–221.

Johnson, Lyndon B. 1965. "Special Message to the Congress on Conservation and Restoration of Natural Beauty." February 8. www.lbjlib.utexas.edu/johnson/archives.hom/speeches.hom/650208.asp.

Korte, Greg. 2013. "Obama's Point Man on the Environment." *USA Today* (April 2).

Lewis, Michael. 2018. *The Fifth Risk.* New York: Norton.

McConnell, Mitch. 2015. "States Should Reject Obama Mandate." *Lexington Herald-Leader* (March 3).

McKibben, Bill. 1989. *The End of Nature.* New York: Random House.

Mooney, Christopher. 2005, 2006. *The Republican War on Science.* New York: Basic Books.

Mooney, Christopher. 2017. "CDC Gets List of 'Forbidden Words.'" *Washington Post* (June 17).

Mooney, Christopher and Jason Samenow. 2017. "Ocean Science Agency Chief Floats Removing 'Climate' from Mission Statement." *Washington Post* (June 25).

Murray, Robert. 2017. "Letter to Vice President Michael Pence" (March 1). www.nytim es.com/interactive/2018/01/09/climate/document-Murray-Energy-Action-Plan.html.

Pelzer, Jeremy. 2009. "Smart-Grid Stimulus Funds Come to Wyo." *Casper Star-Tribune* (October 27).

Rabe, Barry. 2004. *Statehouse and Greenhouse*. Washington, DC: Brookings.

Ramirez, Cindy. 2013. "94,000 Solar Panels to Help Power Part of Fort Bliss." *El Paso Times* (April 6).

Rogers, Paul. 2019. "How California Has Blocked Trump's Environmental Rollbacks." *San Jose Mercury News* (July 27).

Rozell, Mark J. and Mitchel A. Sollenberger. 2009. "Executive Privilege and the Unitary Executive Theory in the George W. Bush Administration" in *Rivals for Power: Presidential-Congressional Relations*, edited by James A. Thurber. Lanham, MD: Rowman and Littlefield.

Schoonover, Rod. 2019. Op-ed: "The White House Blocked My Report on Climate Change." *New York Times* (July 30).

Sengupta, Somini. 2017. "As Trump Exits Climate Agreement, Other Nations are Defiant." *New York Times* (June 1).

Shachtman, Noah. 2012. "Republicans Order Navy to Quit Buying Biofuel." *Wired* (May 14).

Shafie, David M. 2011. "The Presidency and Domestic Policy" in *New Directions in the American Presidency*, edited by Lori Cox Han. New York: Routledge: 166–179.

Shogren, Elizabeth. 2011. "Air Force and Navy Turn to Biofuels." *NPR.org* (September 22).

Strassel, Kim. 2009. "The EPA's Carbon Bomb Fizzles." *Wall Street Journal* (December 11).

Stratton, Allegra, and John Vidal. 2009. "Hedegaard Resigns as President of Copenhagen Climate Summit." *Guardian*. December 16.

Thaler, Richard H. and Cass R. Sunstein. 2008, 2009. *Nudge: Improving Decisions about Health, Wealth and Happiness*. New York: Penguin.

Trump, Donald J. 2017. Executive Order 13783, "Promoting Energy Independence and Economic Growth" (March 28). www.whitehouse.gov/presidential-actions/presidentia l-executive-order-promoting-energy-independence-economic-growth/.

U.S. Department of Agriculture (USDA). 2012. "Vilsack, Mabus Highlight Navy Energy Independence in Norfolk." News Release. September 10.

U.S. Department of Energy (DOE). 2011. "Energy Secretary Chu Announces Five Million Smart Meters Installed Nationwide as Part of Grid Modernization Effort." News Release, June 13. http://energy.gov/articles/energy-secretary-chu -announces-five-m illion-smart-meters-installed-nationwide-part-grid.

U.S. Environmental Protection Agency. n.d. "Federal Green Challenge." www.epa.gov/ Region09/federalgreenchallenge/spotlight/capital.html.

U.S. Global Change Research Program. "Fourth National Climate Assessment." https:// nca2018.globalchange.gov.

USEPA. Administrator Lisa P. Jackson. Remarks to the United Nations Climate Change Conference in Copenhagen (December 9, 2009).

Victor, David. 2011. *Global Warming Gridlock*. Cambridge: Cambridge University Press.

Walsh, Kenneth. 2009. "Carol Browner on Climate Change." *US News* (March 9). www. usnews.com/news/energy/articles/2009/03/09/on-climate-change-the-science-has-just-become-incredibly-clear.

Waxman, Henry. 2008. Opening Statement of House Committee on Oversight and Government Reform, Business Meeting Regarding Contempt Resolution (June 20). One Hundred Tenth Congress.

White House. 2015. "US-China Joint Presidential Statement on Climate Change." News Release (September 25).

Whitman, Christine Todd. 2004. *It's My Party Too*. New York: Penguin.

7

ENVIRONMENTAL PROGRESS AND REGRESS

Following the 2016 election, Republicans found themselves in a position they had not enjoyed in decades. The party held majorities in both houses of Congress, and with the victory of Donald Trump, they had an unobstructed path to act on a legislative agenda.

Many of the people in power had been vocal critics of the nation's environmental statutes, including NEPA, the Endangered Species Act, and major provisions of the Clean Air Act and Clean Water Act. For nearly 50 years, opponents had not been able to dismantle the environmental protection regime established by the nation's environmental laws. Despite unified control of the federal government, Trump and the Republicans never considered repealing these landmark statutes. It was not necessary. Instead, Trump pursued an agenda of environmental policy retrenchment which relied on the instruments of the presidency.

Congress played a minor role in this deregulation campaign, and not necessarily a legislative one. The Congressional Review Act (CRA) was a 20-year-old law created to rein in excessive uses of executive authority, and it was used to overturn a handful of environmental regulations from the final year of the Obama Administration. Every vote was taken with the blessing of the White House, followed by signing ceremonies in the Oval Office. Except for those CRA resolutions, Congress was not needed to implement Trump's agenda of environmental deregulation.

In conditions of gridlock, new legislation is not required to advance an agenda of environmental policy retrenchment. Beyond repealing individual regulations, the Trump Administration imposed restrictions on the volume of rulemaking, marginalized the role of science, and changed the math of regulatory analysis. This assault on the regulatory state may reduce its capacity to advance environmental protection policies in the future. As Judith Layzer observed, the most effective retrenchment strategies were those which operated outside the spotlight of Congressional politics. These include centralized control over rulemaking, regulatory review, voluntary alternatives to traditional regulation, and devolving some policies to the states. All of these, according to Layzer, are pieces of a long-term project of retrenchment.[1]

From the Administrative Presidency to Presidential Administration

The White House has been a prominent actor in the environmental policy arena since the beginning of the environmental era. Beginning with Nixon in the 1970s, presidents have gradually centralized the administrative state, giving them greater control over policy implementation. What changed around 1992 is the central role the presidency assumed when by stepping into the void left by Congressional inaction. As Mary Graham writes, the EPA in the 1990s attempted to reform the major environmental policies of the 1970s without waiting for Congress.[2] Acting with the blessing and support of the White House, environment and natural resource agencies have been able to affect change on their own initiative, in spite of the deadlocked politics that preclude legislative reform.

In the current climate of partisan rivalry, deadlocked politics have precluded the legislative process from delivering effective policies to address longstanding environmental problems. By using the regulatory state as an extension of the White House policy establishment, recent presidents have been able to overcome political deadlocks and make policy changes through the administrative process. Because the environmental policy landscape has changed little over the past 20 years, the legislative branch has allowed presidents to take control of the bureaucracy and influence policy. The safest and most efficient way for the White House to affect policy change is by wielding the instruments of executive power to push forward despite the logjam.

Every recent president has developed institutional capacity and strategies to expand White House influence over environmental policy, as the previous case studies illustrate. Each has availed himself of the same strategies developed by his predecessors. Indeed, a comparison of the regulatory process under Clinton and George W. Bush found that the two presidents used remarkably similar strategies, albeit to different ends.[3] This is a consequence of the centralization of the presidency that occurred and evolved over time, as presidents determined that policy success depended upon the willingness to assert control over a vast and fragmented executive branch.[4] To a large extent, this is also a consequence of gridlock in the legislative branch. A political equilibrium since the early 1990s has precluded any major revisions, reform or repeal of the nation's environmental statutes. Old statutory requirements remain on the books, and this "logjam" exacerbates the problems of conflicting mandates.[5] Under these conditions, presidential leadership is necessary (but not always sufficient) to move forward.

From the Clinton era to the Trump administration, these cases have analyzed aspects of this involvement in policy making in the areas of land management, water pollution, toxics, and climate change. The case studies have demonstrated a developing trend over the past 20 years of what may be termed the environmental presidency. After all, environmental policy gridlock has proved a tall hurdle to clear for administrations since the favorable public opinion and Congressional activism of the 1970s. However, since the 1970s, the environmental community has become well positioned to interact with the executive branch, and some administrations and agencies have found this growing direct interaction very effective.

The most successful environmental policy initiatives to control pollution and manage natural resources and land conservation have originated with the administration. Each administration faced challenges of legislative gridlock, partisan dissention, the legacies of preceding presidents, public opinion and competing policy priorities. Primary tools used to confront such challenges were: executive orders, administrative discretion, directives, rulemaking, and other creative administrative strategies. Whatever the challenge, presidents had to turn to more creative strategies in executive leadership due to the congressional gridlock and other forms of opposition. Of course, trade-offs with other agenda items and priorities have long been part of the executive toolbox as well. Obama used all of the tools at his disposal to redirect previous administrative policies, such as when the EPA immediately withdrew Bush-era rules that relaxed reporting requirements for oil refineries and revoked the permit for the largest planned mountaintop removal coal mine.

This study has explored the evolution, the advances, and the retreats of three recent administrations in order to provide a new approach to and analyses of the role of the presidency as a central actor in environmental policymaking. The case studies in the preceding chapters have followed the pendulum swing of the environmental presidency, as it swung back and forth with rules made then overturned, as regulation and de-regulation altered the contours of environmental policy. The White House's involvement in environmental policymaking and rulemaking in the face of such impenetrable gridlock and other challenges has gone from the advent of the EPA ad congressional activism, from increased executive rulemaking or de-regulation to being supportive of a new progressive federalism. When the legislative processes is bogged down, the environmental president must step in to the fray or develop new executive strategies to bypass stalemates in the other branches of government, strategies that range from overseeing regulatory development, to directing government purchasing and beyond. Building institutional capacity and choice of appointees is vital, as it the strategic use of the Solicitor General to defend administration policies from legal challenges.

Executive rulemaking that was both assertive and creative led to innovation in environmental policy, particularly more recently in the case of climate politics. After the environmental decade of the 1970s, the Reagan administration favored deregulation particularly with the environment, overturning rules and rolling back conservation policies. Due to overreach, and a subsequent backlash, Regan's anti-regulatory agenda stalled in his first term. The institutions and administrative strategies developed by Reagan were put to use by the Clinton administration to bypass legislative channels and serve the cause of environmental regulation. Because public concern for the environment still lacked the intensity of the environmental era of the 1970s, Congress was reluctant to pursue a strong environmental agenda, and most policy change occurred as the result of executive leadership.

Part of Clinton's strategy was empowering career rank-and-file members of his administration. His National Performance Review led to significant reforms within federal agencies, making government more efficient and responsive to the public.[6] But it also gave those government workers a sense of efficacy by inviting them to identify outdated and unnecessary regulations and giving them a say in the direction of their agencies. This approach would later be taken up by President Obama. On a January afternoon in 2012, Obama paid a visit to the EPA in Washington, DC to praise the staff. "Our country is stronger because of you. Our future is brighter because of you. And I want you to know that you've got a president who is grateful for your work and will stand with you every inch of the way as you carry out your mission," he told a room of nearly 200 employees.[7] His hopes of climate legislation dashed early in his presidency, Obama also commiserated with the workers, who had been bashed relentlessly by many of the same members of Congress who obstructed the president's legislative agenda. The pep-talk showed that he appreciated the agency staff and understood the pressures they endured. In bonding with agency staff, Obama was cultivating a valuable resource for his presidency.

George W. Bush used the tools of the administrative presidency to serve an agenda of deregulation. The Bush White House maintained centralized control over the regulatory process. OMB and OIRA to ensure administrative rules were justified from a cost-benefit perspective, such as new emissions standards for diesel trucks. The focus was mainly on the front-end of the regulatory process. Outside of certain areas of interest, especially energy policy, the White House played a minimal role in agency rulemaking. This centralized oversight of regulatory analysis continued under Obama, but to different ends. Agencies were directed to include social costs and benefits in their analyses, and when experts agreed the price of carbon should be set at $42/ton, it provided the justification needed to pursue aggressive GHG reduction policies.

Under conditions of gridlock, the Obama cabinet pursued an executive-centered policy agenda. Obama had to contend with partisan opposition to his agenda, but also pressure to expand ecologically risky oil projects, including the controversial Keystone XL pipeline from the Canadian Tar Sands and offshore drilling on the outercontinental shelf and in the Arctic. Thus, Obama had to balance these energy goals with development of renewable sources, take executive action to reduce pollution, and also address the consequences of climate change. During his presidency, the EPA's mandate was expanded to address climate change and further reduce pollution from stationary and mobile sources. Even with the advances in research and development of sustainable energy that made a relatively "green" president, the context for Obama's environmental challenges was nowhere near as favorable as the 1970s pro-environmental public optimism had been. Obama's environmental presidency also results as part of and has consequences for recovery. Funding from the 2009 Recovery and Reinvestment Act also gave energy and environmental policies a boost, as it became part of the White House environmental agenda. Public investments in research and development for renewable energy during his first term dried up, but many of those programs served as incubators for the growth of private industries. Moreover, energy conservation programs that began with federal stimulus funds were continued by state and local governments.

The move toward progressive federalism became clear with Obama's reversal of the previous administration's opposition to California's waiver under the Clean Air Act. This signaled that California and other states were free to set their own limits on greenhouse gases from mobile sources, representing a shift in the delicate and often acrimonious relationship between the federal government and the states over environmental regulation. Because the Bush administration stood staunchly in the way of states that attempted to put these stricter standards into effect, and then Obama reversed that trend, it will be up to the next administration to find acceptable and effective approaches involving the development of this new progressive federalism as well as other creative strategies of the environmental presidencies to come.

In addition to progressive federalism, the Obama Administration demonstrated a willingness to make greater use of executive authority than previous administrations. Moreover, increased use of Government Purchasing Power to further environmental goals may continue, as with Obama's use of Recovery funds toward research, development, implementation, and purchasing of alternative technologies of sustainable energy. Purchasing policy is a tool with a lot of torque for the environmental president, particularly useful to affect environmental change. The tool enables the administration to re-direct purchasing in the military and elsewhere. For example, Obama directed the Secretary of the Navy to do a bulk purchase of biofuels for Navy ships, as part of the administration's priority to stimulate growth in market for biofuels. Purchasing may thus become an increasingly important part of the president's arsenal. Leading by example through procurement, the president may thereby support environmentalism by using the purchasing power of the U.S. government, as in such instances. Similarly, Obama directed the Air Force to purchase solar panels to be installed on to provide power to hangars. Use of such approaches from the White House toolbox is the best way to affect change in the next frontiers of the environment and sustainable energy, this in contrast to George W. Bush's previous industry-friendly policies, as discussed above. Moreover, encouraging states to take the lead with progressive federalism or regulatory federalism is giving states latitude to create stricter environmental standards to and methods to enforce standards for agencies, or even to go further than the federal government with standards when they are willing to do so or when it is in their best interest.

Climate policy presents the area of the most vivid example of executive initiative forced by legislative gridlock. Climate change is unlike the first-generation environmental problems of pollution and land conservation that provoked widespread public response throughout the 1970s. To many lawmakers and their constituents, the environmental implications of climate change are still perceived as highly technical, the consequences distant. Legislative branch response to climate politics and policymaking was limited from the beginning, following business countermobilization and anti-regulatory trends of the 1980s. In the early 1990s, George H.W. Bush's participation in the Rio Summit paved the way for the executive branch's involvement in climate policy, showing executive action as the most powerful form of intervention in this burgeoning area of concern. A few years later, all too familiar Congressional gridlock prevented Clinton from pushing through greenhouse gas legislation while at the same time the Senate failed to ratify the Kyoto treaty. Clinton bypassed Congress to reaffirm U.S. commitment to Kyoto in the form of an executive agreement.

The problem with this approach was that George W. Bush could unilaterally withdraw from Kyoto. Bush catered to industry by having agency appointees stall development of greenhouse gases rules that could potentially threaten fossil fuel-derived energy production. In this case, the heavy-handed use of executive leadership had the result of burying of agency reports and suppression of significant findings on the effects of climate change that might prove unfavorable to the interests of industry. As analyzed above, with the case of greenhouse gas policy, states' leadership thus became more important than federal legislation or White House involvement, due to Bush administration's strategies of preemptive federalism.

Obama changed course and supported the state-level climate initiatives that the previous administration discouraged. Both George W. Bush and Obama pursued expansive energy and resource development policies, though with very different agendas and to different policy ends. When Obama quickly revoked Bush's policy toward federalism with an executive order in 2009, returning the rule-making process to the EPA, Congress was still as powerless to legislate on greenhouse gases as it had been back in 1999. The Obama administration also used new fuel economy standards, energy efficiency regulations and purchasing (to promote government use of sustainable fuels, clean energy technology, and stricter standards). Recovery Act spending on sustainable and clean energy and tax credits also contributed to climate progress. Even though a sluggish economy slowed the growth of carbon emissions, the creative use of executive authority is helping the country reach Obama's stated goal of reducing emissions by 17 percent by 2020. Even with some disappointments with Copenhagen to achieve much hoped-for worldwide and federal climate policy change, Obama still endeavored to show the global community that he could better achieve climate goals administratively. More aggressive pollution control policies, including the Clean Power Plan, allowed Obama to make a bilateral deal with China to reduce GHG emissions, and ultimately an executive agreement to join the Paris Agreement.

Climate policy also illustrates the provisional nature of administrative policy-making. Trump unilaterally rescinded Obama's commitment to the Paris Agreement and was able to effectively rescind the Clean Power Plan by ordering changes to the regulatory analysis behind the policy. These were significant setbacks because they were effectively the first GHG control policies to take effect in the U.S., and Trump was able to undermine them using the instruments of the administrative presidency.

The environmental presidency continues to evolve. The legacy of Reagan's institutional presidency became a powerful apparatus in Clinton's hands. He advanced an environmental agenda, virtually undaunted by the many Congressional assaults on the regulatory state. Instead of relying on legislative processes during his time in the White House, Clinton achieved environmental policy changes through rulemaking, further centralized the regulatory review process, reoriented major policies through administrative reform initiatives, and built upon public-private partnerships. Finally, Clinton made aggressive use of unilateral action in the form of proclamations, executive orders, signing statements, and executive agreements. Indeed, Clinton's 1994 E.O. 12898 eventually engendered the EPA's Plan for Environmental Justice, particularly the rulemaking part of the plan, which includes the following three major Environmental Justice strategies for 2014: "Finalize the Interim Guidance on Considering Environmental Justice During the Development of an Action. Facilitate and monitor implementation of guidance on incorporating environmental justice into rulemaking. Develop technical guidance on how to conduct environmental justice assessments of rulemaking activities."[8]

The executive strategies such as those employed by Clinton were to prove instrumental to overcoming regulatory obstacles to Bush's policy agenda. Bush also bypassed legislation, but to roll back existing environmental policies, to relax regulations and to expand existing voluntary agreements and public-private partnerships begun under his predecessor. The Bush administration settled lawsuits against natural resource agencies, and those settlements in turn led to more industry-friendly regulation. Bush's policy-by-rulemaking approach gave the OMB a more active role, in part by including preemption clauses in the preambles of regulations. This discouraged states from responding with stricter standards and minimized the risk of litigation over federalism disputes. Obama would confront this approach and subsequently choose to encourage and empower states.

Obama's environmental agenda was best served by his appointees and by the regulatory process, centralized now more than ever before in the White House, from unilateral actions that have helped avoid congressional gridlock to executive agreements as an alternative to treaty ratification. He also turned to partnerships and voluntary agreements to pursue collaborative policies on pollution and conservation. Just as Bush made every advantage of the tools bequeathed to him by his predecessor, Obama's hand was strengthened by the precedents of the Bush years. Creatively appropriating the strategies of his predecessor to different ends, the Obama Administration settled litigation and used those deals as starting points to initiate new procedures to reverse Bush-era rules, and promulgated new regulations to advance policy goals that had languished during the voluntary experiments of the two previous presidents. With new progressive federalisms and regulatory federalism, alternative energy mandates and sustainability and efficiency standards increasingly have the potential to become the domain of the states.

Trump's strategy of deconstructing the administrative state involved suspending new regulations, changing the cost-benefit calculus to scale back the implementation of existing regulations, and reducing the federal government's capacity to create new ones. The first stage of this agenda involved suspending, delaying and rescinding regulations. These efforts were largely unsuccessful, because the rules had been mandated by law and written according to legally defined procedures. When challenged, the courts would usually rule against the administration. The second phase of his agenda is likely to have lasting effects. The administration has slowed the pace of environmental rulemaking by reorganizing and replacing science advisory boards. Presidential directives limiting the number of regulations and the way regulatory science is conducted are also likely to undermine environmental protection policies going forward. Finally, presidential appointees could have long-term effects on environmental policy. This is especially true for judges, who may reconsider long-held legal doctrines such as deference to agencies.

The Essential Role of the Courts

In each of the cases examined here, the federal courts played an active role by intervening in regulatory decisions and redirecting agency priorities. Because of the paralysis that beset Congress, conflict is displaced, and interests frequently look to the courts to resolve policy questions. Courts have been conducive to this role in the current gridlocked political environment, and this has further shaped the relationship between the White House and regulatory agencies

The federal judiciary serves as a check on administrative discretion, and agencies are frequently rebuffed for failing to adequately implement ENR policies. The most noteworthy example of the judiciary's assertive posture in recent years is the Supreme Court's decision in *Massachusetts v. EPA*. That five-to-four ruling not only determined that the 1990 Clean Air Act authorized the EPA to regulate carbon emissions, but it ordered the agency to justify its refusal to use that authority. In a different kind of rebuke, the D.C. Circuit Court of Appeals struck down the EPA's Clean Air Mercury Rule as insufficient because it violated the law's provision that toxic pollutants must be regulated at the source. Even though the courts overturned agency decisions directed by the White House, the rulings were noteworthy because they returned both matters to the EPA. Thus, even in rejecting decisions made by Bush Administration appointees, the courts avoided opportunities to insert themselves into the policy debate any further.

In contrast with air pollution and climate policy, the influence of the courts over water pollution has been more problematic. Recent decisions have reduced the jurisdictional scope of the Clean Water Act, creating confusion over the law and offering the EPA little guidance for clarification. In other policy areas, the impact of the courts has been mixed. Federal courts at the district and appellate levels issued rulings on the Forest Service's Roadless Rule after the Bush Administration chose not to defend it against opponents' litigation, leading to a confusing labyrinth of conflicting rulings rather than clarification. With Congress unable to resolve the conflict through legislation, persistent questions about national forest road construction and maintenance continue to be addressed through rulemaking.

On balance, the federal courts have been more assertive when agencies are influenced by the White House to issue rules that fall short of statutory mandates, and to shirk their responsibilities altogether, as the Bush Administration was alleged to have done. This suggests an administrative strategy for policy retrenchment is more likely to be challenged with vigorous judicial review. Given the significant role of the federal courts in reviewing administrative policymaking, it is very likely that judicial nominations will become more politicized in the future. In an indication this is already underway, Senate Republicans pursued what was labeled a "reverse court-packing scheme." They refused confirmation hearings for three Obama nominees for the U.S. Court of Appeals D. C., holding them up well into his second term.

The case studies also reveal practical limitations of the administrative strategy. Bush's reversal on Kyoto occurred during his first year in office, at the same time the decision was made to suspend the arsenic standard for drinking water discussed in Chapter 4. Both might have seemed like politically safe moves at the time. Noting a declining salience of environmental issues since the early 1990s, the administration gambled on a minor backlash. A Pew Center survey from early 2001 found that just 20 percent of the public knew about the decision to withdraw the arsenic standard, even though 57 percent said they disagreed with the decision when informed about it.[9] The numbers are similar to public opinion on the Kyoto treaty. Only 20 percent reported that they knew about the administration's decision to withdraw, but in that case only 47 percent of respondents disagreed with the move when it was explained to them. The arsenic decision was little-known to the public, but unlike Kyoto, it proved to be very unpopular in the aftermath of the move.

Summary and Conclusion

As the cases in this volume demonstrate, administrative policymaking is a strategy to cope with the deadlocked politics of the legislative arena. Although it is not a cure, it can be an effective method for managing a chronic condition. It has become a useful, if unpredictable strategy in the current state of legislative stasis, which took hold after the environmental era of the 1970s. During that era, there was heightened awareness of environmental problems, such as air and water pollution and land use and conservation, with a public and a legislative branch that were largely motivated to bring about environmental change. This decade also saw agency restructuring and the growth of activism. In another way, the 1980s trend toward deregulation also motivated a backlash response from environmental groups, who were forced to play defense after a decade on the offense, with a voice loud enough to be heard and acted upon by Congress. Business and industry formed groups to enter the fray at this point as well. Over these two decades, the executive branch did not need to develop as many leadership and rulemaking strategies to push through environmental rules and standards as would Clinton, Bush, and Obama in the decades to follow.

As the floodlights that shone so brightly on environmental concerns in the 1970s and 1980s faded, Congress would reach a stalemate, and without the public or legislative spotlights on environmental problems, it became more difficult for the White House to be an agent of change to the environment. Activist groups too adapted and found access points besides Congress to further their causes. The case studies have shown how administrations have created strategies to deal with this stalemate. Thus, with the growth of the Administrative Presidency, presidents over the past two decades chose to bypass much of the heavily congested legislative and judicial routes familiar in the 1970s and 1980s in favor of unilateral actions to further their environmental agendas. The 1990s were characterized in part by a role reversal of the executive and legislative branches, as effective policymaking strategies of the administrative presidency continued to emerge.

An effective environmental presidency is one that makes use of the tools and strategies of the administrative and institutional presidency. In addition to unilateral actions, regulatory review has been a powerful presidential strategy for both parties, from the institutionalized Reagan White House on, as analyzed in the case study of the second chapter. Regulatory review has been particularly effective because it allows the president, acting as chief executive, to oversee and affect many steps of the policymaking process; it also allows the White House to centralize review and policy initiative and to thereby move more quickly through the process. Here the administrative president may make use of bureaucracy rather than being held back by it. Clinton, Bush, and Obama all moved decisively acting as chief executives, putting to use regulatory review to simultaneously bypass Congress and to direct policy initiatives. Both presidents avoided legislation and litigation, preferring to advance their agendas administratively.

The Administrative Presidency of Donald Trump used the tools of the office not just to relax environmental regulations, but to undermine them by immobilizing regulatory agencies. His campaign to "deconstruct the regulatory state" managed to stall and stymie policy implementation without repealing or rewriting the nation's environmental statutes. The Trump presidency began with a governing majority in Congress, yet he preferred unilateral and prerogative actions.

Policy-by-rulemaking strategies proved effective for presidents with very different environmental policy goals. An evolution from the 1970s to the Obama White House is apparent. First, moving from accusations of preemption to a new progressive federalism, the Obama White House empowered states to set stricter standards when they choose. These tactics are consistent with the centralized environmental policymaking coming from the White House. Second, new strategies of unilateral action and use of contextual tools have been developed, and the growing institutional presidency has further enabled such strategies.

In the short run, administrative strategies allow presidents to bypass a deadlocked Congress and address urgent environmental problems. Because lawmakers have been slow to reform the nation's environmental statutes, administrative agencies are often torn between conflicting mandates and using their discretion to set priorities. Thus, the president is in a unique position to give agencies direction. Furthermore, due to the nationwide constituency, presidents can be more responsive and accountable, whether they campaign for stricter enforcement of pollution standards, or relaxed regulations on energy production. Despite the advantages of an administrative strategy, there are negative long-term consequences. To the extent presidents and their appointees can make critical policy decisions on their own, it becomes less likely that Congress rise to the challenge of addressing simmering conflicts that can only be settled rough the legislative process. The administrative presidency is likely to continue to occupy the nexus of environment and natural resource policymaking, and whether this trend is constructive or corrosive depends upon who wins the White House, and the choices made by the individual and inner circle that takes charge of the apparatus of the bureaucracy.

Notes

1 Judith A. Layzer, 2012, *Open for Business: Conservatives' Opposition to Environmental Regulation*, Cambridge, MA: MIT Press.
2 Mary Graham, 1999, *The Morning After Earth Day*, Washington, DC: Brookings.
3 Stuart Shapiro, 2007, "Presidents and Process: A Comparison of the Regulatory Process Under the Clinton and Bush (43) Administrations," *Journal of Law and Politics* 23(2): 393–418.
4 Terry M. Moe, 1985, "The Politicized Presidency" in *New Directions in American Politics*, edited by John E. Chubb and Paul E. Peterson, Washington, DC: Brookings.
5 David Schoenbrod, Richard B. Stewart and Katrina M. Wyman, 2010, *Breaking the Logjam*, New Haven: Yale University Press.
6 National Performance Review, 1995, "Creating a Government that Works Better and Costs Less," U.S. Government Printing Office.

7 Barack Obama, 2012, "Remarks at the Environmental Protection Agency" (January 10), online by Gerhard Peters and John T. Woolley, The American Presidency Project, www.presidency.ucsb.edu/ws/?pid=98862.
8 USEPA, 2011, "Memorandum of Understanding on Executive Order 12898," www. epa.gov/sites/production/files/2015-02/documents/ej-mou-2011-08.pdf.
9 Pew Research Center, 2001, "Tepid Opposition from Democrats, Little Fallout on Environment" (April 26), http://people-press.org/2001/04/26/bushs-base-backs-him -to-the-hilt.

Bibliography

Graham, Mary. 1999. *The Morning After Earth Day*. Washington, DC: Brookings.
Layzer, Judith A. 2012. *Open for Business: Conservatives' Opposition to Environmental Regulation*. Cambridge, MA: MIT Press.
Moe, Terry M. 1985. "The Politicized Presidency" in *New Directions in American Politics*, edited by John E. Chubb and Paul E. Peterson. Washington, DC: Brookings.
National Performance Review. 1995. "Creating a Government that Works Better and Costs Less." U.S. Government Printing Office.
Obama, Barack. 2012. "Remarks at the Environmental Protection Agency" (January 10). Online by Gerhard Peters and John T. Woolley, *The American Presidency Project*. www. presidency.ucsb.edu/ws/?pid=98862.
Pew Research Center. 2001. "Tepid Opposition from Democrats, Little Fallout on Environment." (April 26). http://people-press.org/2001/04/26/bushs-base-backs-him -to-the-hilt.
Schoenbrod, David, Richard B. Stewart and Katrina M. Wyman. 2010. *Breaking the Logjam*. New Haven: Yale University Press.
Shapiro, Stuart. 2007. "Presidents and Process: A Comparison of the Regulatory Process Under the Clinton and Bush (43) Administrations." *Journal of Law and Politics* 23(2):393–418.
USEPA. 2011. "Memorandum of Understanding on Executive Order 12898." www.epa. gov/sites/production/files/2015-02/documents/ej-mou-2011-08.pdf.

INDEX